'Sabrina Mahfouz is a tidal wave c banks of empire with a torrent of not be damned. *These Bodies of Water* is so vast, yet achingly intimate. It is a brilliant piece of work which had me hooked from start to finish' **Lemn Sissay**

'A writer of staggering conviction, ingenuity and integrity. Her skills are mighty, her language is beautiful and precise' **Kae Tempest**

'A brilliant and fascinating history of the Middle East, forcing us to rethink everything we thought we knew. A completely necessary book for us all as we sit in a post-Brexit world, once more considering boundaries and border lines. Sabrina writes beautifully, using prose, rhythm and poetry, to combine history and memoir in one of the most compelling journeys through the Middle East' **Salma El-Wardany**

'I'm a huge fan of Sabrina Mahfouz and her unsurpassed talent to draw a reader in from the first line. *These Bodies of Water* is a great example of her integrity, her essential honesty, this work is breathtaking and ferocious, uncompromising and powerful' **Salena Godden**

'A masterful merging of the personal, political and poetic, Sabrina winds her own gripping, infuriating and often very funny story with the politics involving the Middle East that many, myself included, had no knowledge of, and need to know. This book is outstanding' **Hollie McNish**

'Just wonderful . . . Such a brilliant and unique way to tackle the impacts of colonialism on a region. I absolutely loved it' **Alya Mooro**

'Completely compelling, funny, unnerving and relatable while also being eye opening. A really accomplished and moving piece of work' **Hannah Khalil**

'Fierce, intelligent, and wise, and everyone should read it' **Joanna Harris**

'Sabrina Mahfouz's poetic talents come to the forefront in this lyrical meditation on the influence of the British Empire in the Middle East. Part memoir, part history, *These Bodies of Water* defies categorisation in favour of a lucid, tumbling narrative that sweeps you along for the ride. Like all truly brilliant books, it's impossible to put down while you're reading, and impossible to forget about when you've finished' ***Glamour***

'Hugely creative and engaging . . . *These Bodies of Water* is not your typical non-fiction book. Just as with the rest of her work, Sabrina bends genres with a cutting-edge book that is part memoir, part historical non-fiction, part dialogue, and part spoken word and poetry' ***New Arab***

'With fierce honesty and lyricism, Mahfouz maps out the story of British colonialism in the Arab world, grappling with the dark foundations of her privilege and challenging the lie of meritocracy on which modern Britain is built' ***Culture Whisper***

Sabrina Mahfouz is a writer and performer, raised in London and Cairo and working across multiple art forms. Her poetry collection, *How You Might Know Me*, was a 2017 *Guardian* Best Summer Read and she is an essay contributor to the multi-award-winning *The Good Immigrant*. Her previous theatre work includes *A History of Water in the Middle East*, *Chef* and *Dry Ice*. Sabrina has edited the anthologies *Smashing It*, *Poems From a Green and Blue Planet* and *The Things I Would Tell You*.

Further work by Sabrina Mahfouz

As editor

Smashing It: Working Class Artists on Life, Art and Making It Happen

The Things I Would Tell You: British Muslim Women Write

Poems From a Green and Blue Planet

Plays

The Clean Collection

Chef

Layla's Room

Offside

Noughts & Crosses

Plays One: Sabrina Mahfouz

A History of Water in the Middle East

Metamorphoses

Poems

How You Might Know Me

SABRINA MAHFOUZ

THESE BODIES OF WATER

A Personal History of the British Empire in the Middle East

TINDER PRESS

First published in 2022 by Tinder Press
An imprint of HEADLINE PUBLISHING GROUP

This paperback edition published in 2023 by Tinder Press
An imprint of HEADLINE PUBLISHING GROUP

6

Cataloguing in Publication Data is available from the British Library

ISBN 978 1 4722 8250 7

Designed and typeset by EM&EN
Printed and bound in Great Britain by Clays Ltd, Elcograf S.p.A.

Headline's policy is to use papers that are natural, renewable and recyclable products and made from wood grown in well-managed forests and other controlled sources. The logging and manufacturing processes are expected to conform to the environmental regulations of the country of origin.

HEADLINE PUBLISHING GROUP
An Hachette UK Company
Carmelite House
50 Victoria Embankment
London EC4Y 0DZ

www.tinderpress.co.uk
www.headline.co.uk
www.hachette.co.uk

For the diasporic dreamers

and Tabari, always

History is something that happens to other people

Anonymous

There is no history, only fictions of varying degrees of plausibility

Voltaire

Contents

Author's Note

The 'interviewer' in this book is a fictional character amalgamated from fragments of lived experience for narrative purposes. I did work at the Ministry of Defence and I did partake in numerous Developed Vetting interviews, as would anyone requiring high-level governmental clearance, and the questions asked did spark a personal reflection, but those questions are not documented here verbatim. In keeping with the style of the book, I have allowed my imagination to lead from a lived moment and any dialogue or character traits attributed to the interviewer should therefore be read as fictional.

The short story in 'Evaporation (Palestine)' is a fictional work adapted from a BBC Radio 4 *Short Works* piece, 'A Century of Results', which was broadcast in 2018. The character pieces in the 'Permeable (United Arab Emirates)' and 'Confluence (Jordan)' chapters are fictional works adapted from my 2019 play, *A History of Water in the Middle East.*

Prologue

The way this works is how it works.
I ask what I ask and you mustn't think it is to hurt you.
It's to ensure you are who we think you are.
First question then, easy enough –
Tell me about your politics,
specifically your thoughts on British foreign policy,
historically and presently?

~~~

The interrogations began during the frosty mornings of January and continued until the blossoms had blown across Whitehall to sink into the Thames.

I'd been working as a Civil Service Fast Stream graduate for just over a year and I was being promoted to a posting which required me to be allowed 'access to exceptionally secret information'. I had to be vetted, I agreed to be vetted by a 'Developed Vetting Officer', via in-person interviews over an unspecified period of time. I wanted to be promoted, I wanted to progress. I wanted to get to the top of an establishment that had shunned people like me for its entire existence. I didn't want to
~~~

interrogate the reasons why I wanted this or even why it had shunned people like me. It was easier, more proactive, to just get in and get on with it, change things from the inside, be the loud new voice that could no longer be ignored, ruffle feathers with the security of monthly PAYE and a pension scheme.

But watching winter turn to spring behind the head of a posh, white Englishman in the clichéd costume of a belted beige mac and stiff-collar shirt was not easy at all. Every question he asked made me see myself anew, through his eyes. Eyes I'd never had to directly face, with office clothes on anyway, but ones which I'd always been creating myself to be seen by, piece by piece, in one way or another. Knowing my creation would always be a reproduction, never an original, but striving anyway. Covering the cracks, hoping this papier mâché version of me would eventually have enough layers that nobody, not even an interrogator at the centre of recent history's control room, would find their way to the core. Perhaps he didn't. But what he, or at least his questions, definitely did do was spark my own internal interrogation. One which would last much longer than the few months he was granted to make his conclusions as to whether who I was, and who I had been, could be signed off as suitable enough for who I wanted to become.

~~~

I didn't want to be a spy. Not really. I may have wanted to be Jennifer Garner as Agent Bristow in J. J. Abram's ABC action series that began in 2001, *Alias*, but I didn't want to be a real-life, paper-pushing, Western-supremacy-upholding intelligence officer for a United Kingdom that told me through multiple channels of communication that it didn't trust me because I was poor, I was a woman, a potential Islamic extremist and/or harbourer of Islamic extremists, and though I was British, I also came from another country, Egypt. A country seen as colonially inferior yet full of pride and anti-colonial dissent, so therefore suspicious. I didn't want to be a spy, which if I was granted Top Secret Clearance could be one eventual role of many, so what did I want?

I was midway through an MA in International Politics and Diplomacy at London University's SOAS (School of Oriental and African Studies) when I decided to apply to the Fast Stream. The Fast Stream is the UK Civil Service's graduate scheme, which as its name suggests, fast-tracks employees to senior positions across government. It is one of the most popular and competitive
~~~

schemes in the country, with an average of 40,000 applicants applying for fewer than 1,000 jobs per year and annually appearing in *The Times* top 10 of graduate employers. In 2006, the year that I joined, there were 11.1 per cent of successful applicants who came from an ethnic-minority background. This was a 4 per cent rise on the previous year, and I was told many times that I had been part of the first intake which had the informal aim of 'diversified recruitment'. Despite various schemes since then, in 2019 the figure was 13.5 per cent, with zero successful applicants from an Arab background or even an 'Other mixed' background, the two imperfect boxes I would have ticked at the time. Thirteen years later. Zero.

Unfortunately, and counter-productively, there's no intersectional data available, but the last available report shows those from lower socio-economic backgrounds comprised under 4 per cent of successful applicants during the years 2013–16, rising to 7 per cent and 8 per cent for the following two years, a doubling that is still way off the mark, but means around eighty people from this background per year beginning a career path to senior Civil Service rank, rather than the thirty or so out of 1,000 in the years before then. The rise can be attributed to schemes such as 'Diversity Internships' and 'Summer

Programmes', inviting under-represented young people to spend time in a Civil Service department before applying for the Fast Stream. Marketing was also overhauled to reach a wider audience outside of graduate fairs and broadsheet supplements.[1] The figures on this strand of diversity aren't available at all for 2006 and that was long before they had any specific schemes to encourage and enable those from any marginalised background to apply, let alone succeed. In 2006, I was told they were becoming 'more aware' of ensuring their graduates represented a wider mix of universities and backgrounds than usual, but without official targets. I imagine the figure would be no higher than 2 per cent for those of a lower socio-economic background being offered a role for that year's intake, me and perhaps a dozen others. Wherever I'd previously worked and whatever new worlds I'd found myself in, there had always been people from a similar socio-economic background around me; even if it took a while to find them, we would always end up connected somehow, even if only through a knowing nod and smile. I did find those few graduates eventually and one of them remains a great friend today. But he is a man, and white English. To be the only female Arab, the only Middle Easterner, the only mixed Muslim heritage graduate from a lower socio-economic background in a group of 1,000 young people, in my own city, London, was surreal, to say the least. It was such a sudden and vicious shift to find myself existing purely in opposition and it forever changed how I saw myself in my homeland.

A homeland which had afforded me huge benefits and privilege, especially in terms of education. I went to a grammar school for my secondary education, because my primary-school teacher tutored me for the tests and asked my mum to register me for the 11+. I passed the exams only due to the tutoring, because what ten-year-old would know non-verbal reasoning without tutoring? It was a scam, and my primary-school teacher knew that, but she also knew that if she gave me the tools, I would use them, even when nobody wanted me to. I'm aware now of the inequalities exacerbated by selective schools, how the majority of my secondary-school classmates could have afforded private school, and not a single pupil had passed the 11+ without some form of specific tutoring. But, at the time, I felt incredibly, ridiculously lucky to be at a school with such an extensive library and English and Drama teachers who praised and encouraged me endlessly. The social side was a little more fraught, but not in the ways I'd expected my high-school years to be. Going to birthday parties with swimming pools in the garden and walking two hours home because I didn't have bus fare was around the time I began to build up those papier mâché layers. I would laugh it off by saying I wanted to ask passing people for cigarettes and smoke all the way home. It wasn't a total lie; I was smoking five a day by the time I was twelve and, soon after, weed twice a day, bought with money made from shoplifting sweets and stationery I'd sell to the rich kids; and then from my first job selling door-to-door double glazing at thirteen, a job I got from a girl at school who had been the first with the knowing look, the nod, the shared cigarette. The boss used to take us to the pub for lunch and buy us vodka, lime and sodas. He knew how old we were. He said that's why we should always hope for men to answer the door when we knocked, they'd be more likely to invite a nice young girl in for tea and to talk about redoing their windows. I didn't stay long, after a hand on my knee and a knocked-over mug in a kitchen made me search for a job that didn't involve sleazy men. It would be over a decade before that happened, but at least at fourteen years old I could work in retail on the weekends, something I did for the following four years. I wasn't living in poverty, I had food to eat and a bed to sleep in, I was learning practical and essential life skills and now I am grateful for all the working experiences, even though at that time they were seen as a matter for concern by my school, a distraction from study and an indictment of a household that mustn't have been able to understand that. I was only ever bullied for my Arabness, or harassed and assaulted for being a girl, yet the main thing I felt shame about was money, the lack of it and what needed to be done to get it. Deep, hot shame at not knowing anyone that would be called 'a qualified professional', not even within the 'vocational' industries, such as electricians or plumbers. There were students with parents who did this work and made huge amounts of money. Their social standing was far beyond what traditional British class theory could explain and I grew up in awe of them, fathers particularly, who were vocationally trained. It was only when I started university that I realised these professions were regarded as 'traditional working class' and that I had, in many ways, grown up aspiring to fit into a classification which had then seemed so out of reach, even if wider society would have labelled me as this at the time. During my school years, another feeling of shame was the shame for feeling shame, knowing how proud my mum was that she now lived in a middle-class area with a daughter at a grammar school. That shame, it showed me how to code-switch magnificently. That is, I learnt how you must present yourself, or which parts of yourself to present and which to hide, in order to be welcomed in places where your whole or real self would not be welcomed. Drawing on my non-Arab self for post-9/11 travel purposes; imitating the aloof confidence during large social gatherings of the wealthy girls at my school who grew up in their own wing of a house; presenting the most theologically interested side of me to a religious family in order to find their favour; this might sound like regular behaviour humans employ in order to adapt to their changing environments. And it partly is. What makes it specifically code-switching, rather than just an amiable way of 'fitting in' and finding acceptance, is its incessant presence. The way a switch has to happen no matter where you are or who you are with, the exhaustion of never being able to say, this is all of me, to anyone, ever. The fact of your reliance on it to make your temporary taking up of any space tolerated, if not quite welcomed, alerts you to the reality that there is no place of acceptance for you. Of course, due to being able to pass as white, I was certainly able to code-switch in ways that gave me far more access and opportunities than those who could not pass as white, Arab or otherwise. I will never be able to ascertain exactly how many of the opportunities I've had, despite the intersectional marginalisation of my identity, have been down to white privilege, but I can make an informed guess that it is almost all of them. My code-switching was nothing compared to what those of the global majority who can't pass as white have to master for daily survival in Western countries, and in many outside of the West too, including the Middle East. But it did give me the power to move through an otherwise impassable world, and it also saved my life more than once. The switches felt more extreme in some places than others. The estates, the late-night parties, the pubs, the clubs, the clandestine corner drug deals, the chatting shit on park benches, the prison visits, the nightbuses, the overworking until falling asleep in the stockroom seemed so easy for me – in comparison to the pool parties, the middle class dinner table debates, and later the expectations of professors and fellow students – that they became my safety, a meditation, even as they put me in endangering situations regularly. Others would dream about leaving – the retail job, the bubble we created with drink and drugs and raves, the hand-to-mouth of it all. I would dream about both staying and leaving – I wanted one foot in and one foot out. The comfort and the cold. Even when I worked 9–5 in the Fast Stream, I continued working in nightclubs at the weekend. I justified it as necessary to make the money to pay for my MA, which is partly true, but now I see it was also my comfort, whilst the Civil Service was the cold, the place I had to exhaust myself to be tolerated.

I was only able to get accepted in the highly competitive Fast Stream due to my education and the various work experience roles that came from that. The privileges afforded to me by the education system at the time extended to me not having to pay a penny for my undergraduate degree, as students from low-income single-parent households didn't have to in those days. It's worth noting that many countries, including Scotland, Argentina, Greece, Germany, the Czech Republic and, to an extent, India, offer free undergraduate education to all their citizens in an attempt to ensure widespread access to higher education across their populations, an aim that seems so far removed from what England wants to achieve nowadays.

I still had to work to fund my living expenses, but in London, there were countless opportunities for working through the night as a young woman. I took plenty of them. During one of those opportunities, waitressing at Stringfellows stripclub in Covent Garden, my Kosovan colleagues became great friends, and through them I did informal work experience at an office in Pristina that was supporting the United Nations War Crimes Tribunal in Kosovo. It was officially a post-conflict zone then, but one still administered by the UN and full to the brim with international aid-and-development agencies. This was the peak of NGOs (Non-Governmental Organisations), supposedly helping to rebuild people's lives without a political agenda. The Kosovans I knew had little trust in these NGOs. They said they needed their own money to rebuild their lives. The agencies were spending plenty on programmes and gardens, but what good would it do them if their businesses remained closed because they couldn't pay the rent and their customers couldn't afford to buy anything? This viewpoint, alongside seeing the national flags of the countries where each NGO originated pitched up outside every space where they were operating – whether in an office building or a children's playground – was the first time I realised the NGO world was in part another version of colonialism, staking a country's claim in another country through the morally celebrated framework of early 2000s humanitarian aid. I didn't doubt many individuals who made up the organisations had compassionate intentions, but I now realised those intentions came with a political agenda. Still, the work I did there, helping to document the victims of war crimes, though it was often traumatic, felt important and necessary, setting me off on a path towards working in international politics, instead of the other avenues I'd explored with previous work experience placements.

One had been at ITV's Factual department, after I'd emailed them to profess my passion for a career in TV production due to my love of research and storytelling. They placed me with the team finding the next batch of struggling celebrities for *I'm a Celebrity, Get Me Out of Here!* and suffering parents for *Teenagers from Hell.* That 'passion' of mine was quickly extinguished. In college, I'd done an internship with Surrey County Council's Archaeology Department, going on archaeological digs at Roman sites and helping to preserve ancient manuscripts and artefacts. Although everyone seemed surreally posh to me, I loved the intense, silent focus of preservation. The way my shared knowledge of and curiosity about a specific historical period easily overshadowed any questions someone might have about my own history. It inspired me to study a BA in Classical Archaeology. Though I loved the history, I struggled with the highly mathematical aspect of archaeological practices (I eventually got an adult diagnosis for dyscalculia, the 'mathematical dyslexia') and I switched to English Literature and Classics for my second and third years. All of these things – the degrees, the work experience placements, the opportunities to discover and experiment with what I was interested in – were open to me in a way they hadn't been to other members of my family, and to so many other people in the country and outside of it, because I had a British passport, lived in London, was white-passing and had learnt to use the power of code-switching to get me where my background could not. The full potential of humanity's decency, decadence and destructive nature were not hidden from me throughout these years. I wilfully waded my way through them on a daily basis, the choice to do so a sliding scale but there, and a privilege nevertheless – stripclubs, war crimes, ancient torture tools, illuminating works of literature, world-changing community activism.

Considering all of this, how could I not have known what I was applying for and why, when I decided to go through one of the most competitive graduate schemes in the country, to get into the heart of the gargantuan machine that runs this disproportionately rich and powerful nation?

Maybe I did and I'd just overestimated my code-switching abilities. Or maybe I'd overestimated how much I knew who I was. Because we are kept from ourselves even if the world is not kept from us. Knowing ourselves is the most dangerous form of knowledge to the continuance of the status quo. Especially if by knowing ourselves we learn how we were made and why and what for. I say we, meaning all the children of an imperial diaspora, those who grow up in a country that colonised their parents' countries; but I also mean anyone who challenges the accepted framing of British society as being democratic, meritocratic, progressive, freedom-loving and giving, and benevolent. If it is seen as all those things, then those at the top of its society must be deserving of their grotesque wealth and their descendants must be deserving of inheriting it. If it is not adamantly and aggressively seen as these things, if it ever drops the ball long enough to let these qualities not be the story, then it is open to being called despotic, corrupt, nepotistic, regressive, cruel, greedy and violent. And if it is all those things, then those at the top of it are no longer seen as deserving and they will have to go to great, ungainly lengths to keep their grotesque wealth and enable their children to inherit it. It is much easier to just make sure we all stick to the same story. Even as I write this, I am fearful of veering from that same story. What responsibility that burdens me with, what ingratitude I will be accused of, what deep dissatisfaction with a part of myself it indicates, what guilt it causes when I recognise all the benefits I have welcomed and all the bad that has been done to provide me with them.

~~~
~~~

Role-playing, box-ticking and non-verbal reasoning my way through a year of assessment centres, I was tearful as I held my UK Government Fast Stream graduate scheme acceptance letter in my hands. I'd submitted the initial online assignments in a Soho basement internet café after my uni classes were over before running off to work my waitressing job at the stripclub until 4am. The subsequent invitations to the anonymous, plush test centres that followed hadn't filled me with dread. Compared to revising passages of Herodotus in Ancient Greek whilst wearing lacy, itchy underwear and waiting for the next champagne order from someone who was certain to engage me in a blunt conversation about his most urgent sexual fantasies, answering hypothetical questions about how I'd negotiate a resources dispute between claimant states was a rush of air to my smoke-filled lungs.

As I held the letter I was proud of myself, for so many reasons. My dad had dreamt of being a civil servant in his youth, but he had left Egypt and higher education to avoid military conscription and to live in his wife's home country. My mum had been a secretary for an oil company and had met and married my dad in Cairo. She'd left school at fourteen with dreams of one day going to art school, but was forced by her mum into secretarial college and then to full-time work shortly after. So whilst

I was proud of myself, I knew that dreams were just that and I shouldn't get too excited. I folded the letter away and wiped my eyes to begin the make-up routine that marked each evening before stepping down the stairs into Stringfellows' basement and becoming the girl in stockings who brought the bottles. At that moment, my dad was driving a minicab somewhere through London, likely hoping for a vomit-free evening from his customers. We shared that dream too.

My hopes of being accepted onto the Fast Stream were genuine enough that I had worked hard to make it happen, but they were also abstract, a self-test to see how well I could gain access into this solid British establishment, a curiosity to experience the process. Simon Case, the current Cabinet Secretary and head of the Civil Service, was recruited in the same year group as me. I learnt later that he represented how Fast Streamers were expected to approach their placements: strategically. That they should gain experience in finance and project management roles in a particular order to enable them to qualify for the most high profile appointments as their career progressed. This wasn't taught as part of the Fast Stream process, so I assume they are advised by their universities or by their families. It probably seemed like common sense to them. But for me common sense was to ask for a placement that I would enjoy and be able to contribute the most to. Placements weren't assigned at acceptance: you had to reply to the initial letter with your top three choices of department. The National Archives at Kew was my first choice. This is a department created by the merger of the Public Record Office and the Historical Manuscripts Commission. It holds over 1,000 years of official British documents. When I received a phone call in reply to my submission of choices, I was told that of course there were no Fast Stream positions in the National Archives, it was a non-ministerial department, I should be thinking bigger. I was also told that my second choice, the Foreign and Commonwealth Office (FCO), was thinking too big – only the very top percentage of recruits went there and they got personally invited to do so. They were pleased to offer me my third choice, the Ministry of Defence (MOD). They said they'd send an email assigning my role and wished me the best.

I was a signed-up member of Stop the War and had marched chanting 'Not In Our Name' in time to drum'n'bass in Trafalgar Square to protest against the British invasion of Iraq. The anti-capitalist campaign group War on Want sent me a monthly newsletter and that was the most enthusiastic I got about the word 'war'; mostly it gave me panic-filled flashbacks of the burnt-out houses and stacks of mattresses where families were executed that I had to document in Kosovo. Why had I made that particular government department a choice at all, then? I'm still not entirely sure. I know I am drawn to the epic, edged with a need for the unexpected and the opportunity for quantifiable achievement. The National Archives had appealed to me because I imagined being able to piece together untold histories of British colonies, convincing ministers of the need to teach the British Empire more thoroughly in schools with the archives we had at our disposal. If I am being incredibly kind to myself, then perhaps I saw a senior role in archiving British history as a magnificent reversal of soft power, seeing as almost every archive, either ancient or contemporary, detailing both everyday life or monumental battles in Egypt, that I had read was filtered through the eyes of a privileged, European white man. To be able to file, study and disseminate imperial archives seemed to me, at the time, to be one of the most radical things a young woman of mixed Middle Eastern, Guyanese and British heritage could do. Perhaps they thought so too, which is why the request was dismissed instantly.

The possibility of becoming a diplomat brokering world-changing peace deals for the FCO had also appealed to me, which is why it became my second choice. I think the MOD was the only other department which offered a satisfactory sense of the epic and ego. Differences between cultures, countries, languages and religions had sparked clashes throughout much of my childhood, so perhaps there was also some feeling of being best placed at the headquarters of battle, without properly thinking through that unlike at those personal moments, this time I'd have to choose sides – and they'd be armed to the teeth with more than words.

I wrote this and it seemed true. Then I looked through my research papers and I saw my (failed) application to the War Studies Department at King's College dated a year before I submitted that online assignment for the Fast Stream. I just need to admit it: there is war in me, and it isn't pretty.

~~~

War is in all of us, one way or another, a constant part of humanity's long story. Not least, a significant part of the story of every empire, including the British Empire. Britain fought 230 wars during the sixty-four years of Queen Victoria's reign alone, paving the way for its imperial 'peak' of 1921, when it ruled, to varying degrees, 23 per cent of the world's population and 25 per cent of the world's land mass.[2] Britain was not unique in its
~~~

moral justification of these wars. The global moral frameworks to justify war have developed over thousands of years of military violence across the world.

Ancient Egypt had a militaristic ideology that lasted over 3,500 years, with the most expansionist pharaoh, Thutmose III, taking the empire from Nubia across to Palestine and Syria. The Ancient Egyptian kings and queens were quasi-divine, their proximity to the gods giving all the justification to wage war that they needed.[3] This royal ideology still exists, the British crown even now 'ordained by God'. During the expansion of the British Empire, and most others that came before it, the justification of a divine mandate was a standard excuse for the terrible violence required to establish and preserve it. The European colonisation of the Middle East really began nearly 1,000 years ago, with five civilian-slaughtering Crusades led by European Christians between 1099 and 1221. They were initially instigated by Pope Urban II with the ostensible 'just means' objective of expanding the world's access to Catholicism by defeating Islamic rule, which was cleverly inextricable from the real objective of strengthening the Western church's power and wealth by capturing Jerusalem and various territories across what are now known as Palestine, Lebanon and Syria. This reasoning – initiating war on the basis of 'just means' – continues today, depending on what society is willing to accept as just, even if that acceptance hangs by a thread. Most wars are framed as retaliatory, defensive or preventative – they are planning to attack us so we must attack first, or they are stopping us accessing resources we paid for, resources that keep our nation going. The framing of war and its necessity is so intricately woven with all aspects of society that it is difficult to pick apart where it really sits within your own moral framework and how your lived or inherited experience might affect that.

In terms of my experience around the time that I agreed to take a role in the MOD, the impact of the 9/11 terror attacks on the World Trade Center in New York had certainly had an effect on all the frameworks I'd built around me. Like all people with Muslim and/or Arabic names and heritage living in the global north during the 2000s, I was being dismantled and dissected in every situation, every day. In 2003, the Prevent strategy was introduced as UK official policy, as part of the overall post-9/11 counter-terrorism approach called 'CONTEST'. The aim was to prevent the radicalisation of individuals turning them to terrorism. To those of us who were targeted by it, it seemed to regard every Muslim and every Arab in the country as an individual who was susceptible to radicalisation and in need of community interventions to stop us from joining al-Qaeda. I think it's highly probable I internalised this sense of suspicion. If I wanted to reach a senior position in this country, whatever the industry, I would need to prove my non-radicalised status. And what better way than by joining the very forces working against us? It was simply an extension of what I had been doing my whole life, what so many marginalised people from all identities do, change ourselves, our frameworks, in order to assimilate and get to the place we think we want to get to, imagining that once we get there, we can finally be our true selves and stop wearing the masks of assimilated acceptability. A hope that is rarely realised, because once you dismantle yourself for others it is difficult to remember where all the parts once went. There was a 'war on terror' against me in my own country and perhaps I chose the way of the coward, negotiating my immunity by turncoating.

Early on in the vetting interviews for my Top Secret Clearance, once I'd been working at the MOD for a year and needed that clearance to progress to a more senior position, my interrogator asked me about rivers and canals, what I thought about how the British had used them to maintain their empire. Until he asked, I hadn't thought specifically about water and how it surrounded every colonial decision the British had made. How one of the areas in the world I came from was seen as sand and oil, dry and scorched. But their water had enabled the empire, and its legacy that I became a part of, to expand and survive in ways that would have been unthinkable without it. I became quite obsessed with how water forms our bodies and our borders, how it shapes landscapes, lives and legacies, enables and disables empires. All the empires, not just the British one. There have been many. But the British Empire is the one that has viscerally affected my life, positively and negatively, inextricably. The one that continues to affect lives and politics and inequalities across the world.

Water in all its forms is at the centre of our impending environmental breakdown and ecosystem collapse. A world that is not only still at war but refuses to admit its reliance on violence and war for its continued economic survival in its current model makes it difficult to imagine how such a world will be able to deal positively with the uncertain future we face. The sharing of water resources amongst nations offers the hope of collaboration and generosity before greed, but at the same time geopolitical crises such as the dispute between Egypt, Sudan and Ethiopia over Ethiopia's building of contested dams in the River Nile simmer constantly. Almost all conflicts have elements of water politics within them, even if it is not one of the primary causes.

The more research I did, the more it revealed how central the Middle East's water was to this behemoth of a global power, when I'd only ever been told that its interest in the region was to do with oil. Oil has been the main, and most direct, moneymaker, no doubt, but the British dependence on the region's water for fast and easy trading routes goes back almost 100 years before oil was even discovered in the region. By the time it was discovered in Persia, modern-day Iran, in 1908, the British Empire had already colonised, to varying degrees, a quarter of the world, including what are now Egypt, Sudan, Bahrain, Kuwait, Yemen and Oman. This region was growing in importance to the British partly because the Ottoman Empire had begun to dissolve and they wanted to be first in line to gain control of its Middle Eastern territories. The competition between European imperial powers was increasing, and ruling this area allowed Britain unhindered access to the quickest routes for imperial trade and therefore economic dominance. The region was instrumental in getting the British Empire to its so-called 'territorial peak' in 1921, when it ruled, with protectorates, colonies or dominions, sometimes as the East India Company and sometimes as the Crown, over 13 million square miles of the world's land mass, making it the largest empire in history.[4] The main period of British imperial decline and declarations of colonial independence was from the end of the Second World War in 1945 to the 1960s, though it continued right through to the 1990s, when Hong Kong, the last official British colony, was 'handed back' to China in 1997, as per the terms of the 1898 lease China gave to imperial Britain. There are currently fifty-four sovereign states in the Commonwealth, and only two of them, Rwanda and Mozambique, were not part of the British Empire in some way. Queen Elizabeth is still at its head, making it the most explicit continuing legacy of British imperialism, but by no means the only one.

~~~

In this book I focus specifically on water in the Middle Eastern region and the way British imperial exploitation of that region's water rendered the area even more politically complex. I'm not suggesting that this was the main enabler of British colonialism, just one intrinsic facet of its unprecedented expansion. Ideologically, the main enabler would eventually become white supremacy and its required racism. Economically, and inextricably, linked
~~~

to white supremacy, the dominant enablers became slavery and later, to a lesser degree, indentured labour. The colonisation or indirect British rule of the present-day Middle Eastern territories of Bahrain, Cyprus, Egypt, Iraq, Jordan, Kuwait, Oman, Palestine, Qatar, Saudi Arabia, the United Arab Emirates and Yemen may have made it possible for the British Empire to survive as long as it did, but slavery, and later indentured labour, was its fundamental backbone. Every mention of colonialism in this book cannot be detached from colonial slavery and racism. Every page hopes to add whatever tiny amount it can to the constant challenges and struggles against the normalisation of their legacies.

And whilst the Middle East continues to suffer the effects of British colonialism, the area has its own shameful history with regards to imperial expansionism enabled by the enslavement of African people. Military slaves were used by Egypt's rulers for ten centuries, from Ahmad Ibn Tulun (r. 868–84) to the late nineteenth century. The first slave armies were black, the result of Ibn Tulun having trained 40,000 Sudanese slaves.[5] Today, the Sudanese population in Egypt, whether migrants or refugees, are currently facing horrific conditions, including modern slavery and state violence.[6] Anti-black racism is active and accepted across the Arab world, and this has to be acknowledged and challenged by all people from the region calling for a fairer and safer society, something which cannot happen without eradicating anti-black racism. The legacies of British colonialism still disproportionately affect black people, and there are many scholars and activists, particularly from the global south, who have dedicated their professional lives to documenting this and suggesting ways forward, a number of which are included in the reading list from crucial perspectives at the back of this book.

The 'Middle East' is a contentious term. It is a political term, as all borders and regions are, not geographical fact. It is of course 'East' only in relation to 'West', seeing as it is right in the middle of a spherical land mass. The countries that have been labelled as belonging to the region have changed over the years. Afghanistan, Eritrea and Sudan are highlighted in some early-twentieth-century British maps of the Middle East but are rarely included now. Egypt is considered the western cut-off point despite the vast majority of the country being positioned in North Africa. Cyprus is most often culturally positioned as European for political and continuing colonial reasons, despite its geographic location being in the Levant (which is also referred to as the Near East) and sharing sea borders with Israel and Lebanon. Turkey is sometimes included in the Middle East now, though it used to more often be categorised as Asia Minor, and some consider it at least partly European. In American political terminology, Iran is definitely considered the Middle East, but UK commentary generally considers the Middle East as ending at the border of Iraq and Iran, depending on the headline or the point they are trying to prove or disprove. Clearly, the term is outdated, inaccurate, and entirely Anglo-centric and colonial. I continue to use it for this book but look forward to a time when there is an adapted term which comes from the region itself, allowing each country to decide whether it is to be included, rather than being imposed by European and American needs for a classification that aids economic and political hegemonic objectives. I also use it because it is still the mainstream accepted term for the region, but mostly because it reflects so much of what this book is about: how the British have stamped their needs, wants, claims and names all over the region and yet it still feels taboo to talk openly about this; about how it continues to affect so many millions of lives in the Middle East, let alone the lives of those who have become part of or been born in the diaspora. There is so often the sense, even within the diaspora families themselves, that we should not underplay our agency, our luck, and we should not overplay history, because it is just that. But how can history be overplayed? It is in everything that is our present and we should want to know it all, so that we can build the best future.

In this time of climate emergency, we must also be constantly reminded that the British Empire's breadth and reach were only possible due to its being the leading force of the Industrial Revolution. Since that point in the mid-nineteenth century, industrialised countries have generated large quantities of greenhouse gases and they are primarily responsible for human-made climate change. Fragile, often previously colonised regions, pay the price for this, with drought and flooding disproportionately affecting their populations. This is painfully clear across the Middle East, where there is the least amount of renewable water resources per person in the world. Across southern, eastern and central Africa, more than 50 million people suffer from poverty and hunger due to drought and climate change[7] and hundreds of millions of people are affected by devastating and worsening flooding caused by climate change across Asia.[8]

If for no other reason than climate justice and placing greater expectations on those who have historically contributed more to environmental destruction, the positioning of European colonisation as 'moments' of history that no longer affect our present-day crises must be challenged at every opportunity.

Since the first days of those internal interrogations sparked by the Ministry of Defence vetting officer, I've wanted to examine one of these specific 'moments' of British imperial and Middle Eastern history. A moment that lasted from 1914 to 1956. Forty-two years when Britain was the dominant power in the area. Using this power to build the society I would benefit from and to enrich and empower those I would eventually serve, in more ways than one. Displaying little regard for the consequences this would unleash for decades to come, consequences that we can see unravelling on every news feed today.

I wrote and performed a theatre show about this 'moment' for the Royal Court Theatre in London, *A History of Water in the Middle East*. It was highly condensed and very much focused on what the title suggests, within the framing of British imperialism. I chose to look specifically at Bahrain, Egypt, Iraq, Jordan, Kuwait, Palestine, Yemen and the United Arab Emirates through this dual lens of water and empire. I continue to use these countries here. There are many more countries that could be included, such as ones in the region that are even more water-stressed; Qatar currently tops the world list. But each of the eight countries I write about is in the list of the top 20 countries at 'extremely high' or 'high' risk of a water crisis in the near future.[9] I've chosen to highlight women's stories throughout, whether my own, historical or fictional, speculative or contemporary, in a hope that it adds in some tiny way to the redocumenting and reimagining work that must be done to address centuries of global erasure and censorship. Britain has been exceptional in its ability to choose what histories it tells and how it tells them. Whether through school curriculums or big-budget films, the histories that have been told have often been picked, pilfered and filtered to appear as fair portrayals, with mostly male historical figures dealing with personal character flaws rather than in-depth reflections on the uncomfortable realities of the imperialism that has made Britain what it is today.

Whilst there are many more lenses to look through when discussing the Middle East and British colonialism, I hope these by no means exhaustive notes will be enough to encourage people to look them up, to write their own. The forms I flow through change and I hope their relative spontaneity allows a feeling of push-back against the organised way we have become accustomed to absorbing these explorations of unorganised identity. It may just annoy you – as with any of it, skip through as you please.

I am not an academic, an activist or a journalist. I stand in awe and extreme gratitude to them. I also have a concrete connection only to Egypt. That I am given permission and validation to write and represent a region so vast and varied due to that one connection shows how far we have to go. I have never even visited Bahrain, Iraq or Yemen. But I include them here, because I want to attempt to show the colossal breadth of Britain's imperial projects, invasions and exploitations in the region and show how colonisation continues to connect us. Arguably that connection also hinders the act of decolonisation by preventing specificity to be demanded, expected and normalised. These notes must then be personal, specific, spontaneous. I have excavated shoeboxes and USB sticks and have dislodged memories in order to try to make it so. But memory is messy. It is fiction and it is truth, as all fiction and truth are. I have tried to show a side to the British Empire that is less written and spoken about, but it is not the whole of that side, by any means. I have written about my life, but only snippets and not the whole of it, by any means. I have picked, pilfered and filtered which of my histories I wish to tell. I learnt how to do that from the best, after all.

These are my notes on the British Empire, the Middle East and where we meet.

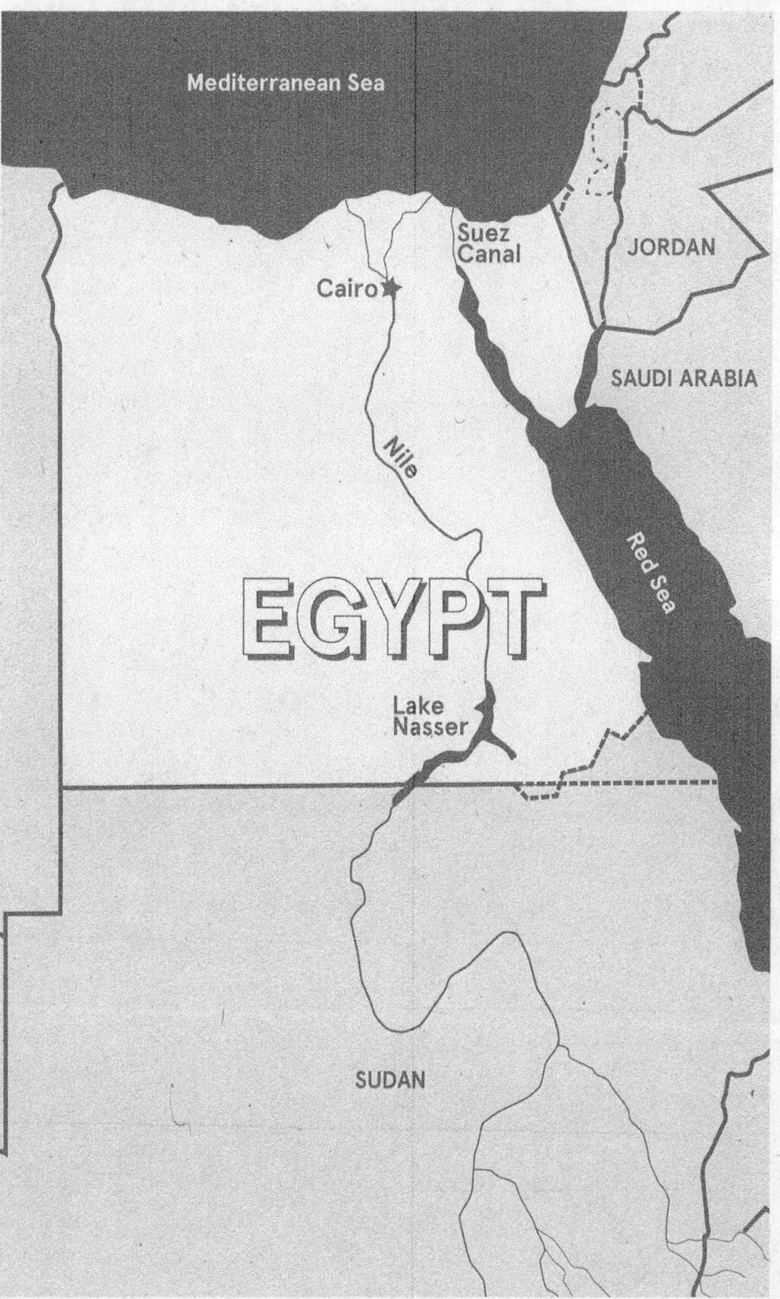

Mediterranean Sea
Suez
Canal
JORDAN
Cairo
SAUDI ARABIA
Nile
Red Sea
EGYPT
Lake
Nasser
SUDAN

Harbour I

(Egypt)

Are you not made of Suez silt?
How do we know you won't
shore our boats
by making yourself bigger
than we made you?

~~~

I was never a sea person. I learnt to swim in a Cairo hotel pool, and after being stung by a jellyfish at a beach by the Red Sea, my conviction that life should be led amongst city towers and car horns and competing muezzins was cemented. The water I loved was city water. Rivers, canals, streams, lakes, puddles.

What does it mean to think of water beyond its being a resource? It is present in the smallest of living cells as well as in gigantic glaciers and oceans. But when I think of what water means to me, it is history, and connection.

Walking along the corniche in Cairo, staring at the feluccas on the River Nile, their sails spread like reversed
~~~

dove wings, was when I first remember being breathless at the history flowing right beside me. Even more so than during my countless visits to the Pyramids, sneaking inside their chambers, giggling with a friend of a friend who knew someone who knew someone. The Nile surpassed everything, and my favourite treat was going to a trampoline park right on the hazy riverbank, jumping higher and higher to get a better view of the fishing boats and the traffic-clogged bridges and the other side, which seemed another world altogether. It felt like everything that had been and that will be, all in one long body of water.

Ever since there has been a telling of history, there has been a telling of Egypt's river. The Blue Nile starting in Ethiopia, the White Nile in Rwanda, meeting in Sudan until it reaches the Mediterranean. Our present human knowledge stream could be thought of as tributaries from the Nile – philosophies, astronomies and mathematics from Kemet, waterwheels, ironworks and furnaces from the empire of Kush. Even the Ancient Romans and Greeks (whose own learning was passed down to us only because of the Arabs) gave the African civilisations credit for all this, but somewhere along the line from Oxbridge to Harvard, white scholars did their best, consciously or not, to decentre the African origins of knowledge. I grew up knowing this at a time before Twitter or You-

Tube, before petitions to teach a broader representation of the British Empire. It's what comes with parents and grandparents from ex-British colonies, their casual way of dusting away the school-history-book propaganda with enough anecdotes to get me awarded a detention at every single history lesson for my 'Nah, miss, it didn't actually happen like that, you know.'

At the time, me and my classmates thought this raucously funny, the history teacher red-faced, skirt swishing with fury at the teenagers who thought they knew the subject she'd studied for decades better than she did. Her glasses being pushed back with one hand whilst her other ripped reams of detention slips from her desk drawer. There are glimpses I can hear first, then see, her smacking a thick textbook down on the desk in front of her, literally frothing at the mouth as she shouts at me and my friends for another of our contradictory interruptions. The friends closest to me all had at least one parent from an ex-colony or protectorate – Mauritius, Hong Kong, Ghana. Whilst we enjoyed provoking our teachers in general, our version of empire was personal, and we knew even then how our histories had been written for us. We did not submit to the perpetual retelling of them, solidifying them for generations to come as something that was palpably untrue – or at the very least a highly edited breakdown of events that left out the most important stories of all, those of the people from those places. In our testy teenage ways, we were attempting to insert those stories that were given to us first-hand from our grandparents or parents into a school curriculum and ethos which had no room for us, unless it was for a prospective brochure photo.

We spent our detentions reading, doodling, note-writing, listening to music. It wasn't a terrible punishment. Busy planning nights out on park benches and, later, what outfit would be most likely to get us into bars and clubs without ID; not much thought was given to the fundamental reasons we were there after almost every history lesson.

Our teacher had probably not given it much thought either. We were disruptive, disrespectful, obstructing other pupils from their learning. Fair enough, we were. But we were also a threat, even aged fourteen, for what we might tell, the way we might tell it. More than timelines of historic events or life stories of heroic figures, in each of those detentions we were slowly learning that there was only one way of seeing history in this country and getting on, undisturbed. To see it differently and to speak about it openly would result in restriction, disfavour and isolation. I got the message, and by the final year I patiently memorised what I was given and loyally regurgitated it to get my History GCSE: A. At the same time, I kept my own history quiet.

~~~

Thirteen days old and I had been a bundle of hand-knitted blankets and booties on a flight to Cairo, Egypt, the city where I'd live until I was two and then again at eight and again at twenty, with a biannual visit whenever it wasn't officially home.

Alhamdulillahs were bestowed upon my pale newborn skin across the districts of Cairo where my dad's family lived. As an older child, I was told that I must know I had been blessed by Allah with my pale skin and light hair; 'an angel, an angel' was shouted across tables as I ate next to my cousins, their dark-brown arms nudging mine, eyebrows raised with 'Here we go again, oh blessed one.'

I may not have looked much like my Egyptian family, but I was always at home in Cairo; it felt like my city almost more than London, the loud, neon chaos spurring me on rather than weighing me down in the way London's comparative orderliness sometimes did. But each year I grew more cognisant that much of this ease was a privilege given with skin shade and all the colonially adjacent status it suggested and, by suggestion, granted.
~~~

For a long time I accepted it, not unquestioningly, but without real interrogation or demands for myself or others to recognise it. Being part Arab had never given me anything but sneers, slurs and suspicion in England, so I wasn't about to reject the flimsy awe with which Egypt often regarded me, despite knowing it to be the other side of the coin that made England belittle me – a coin that was minted by the British Empire and would prove very difficult to bury.

~~~

When I wasn't living there I still travelled to Egypt around twice a year throughout my childhood. My school memorisation of a proud British history with no mention of the violence and devastating legacies of colonialism was rebutted by stories and people who had lived through a very different version of events. I had family members who had felt such shame at their Egyptian dialect they'd sat by a radio for years, repeating BBC World Service programmes until they could speak RP English without ever having left the continent of Africa. Who when they had finally arrived for a holiday in Europe and then, triumphantly, in England, the place of all the cultural capital they cared for, they were racially abused, called the N-word, something they'd never heard before
~~~

and initially smiled at and waved in response to, thinking it was a gruff British greeting. Still, the love remained for this rainy country that had The Beatles and The Rolling Stones, and despite the majesty of the Nile, my Egyptian family would speak of the Thames as though it was the longest river in the world.

~~~

I often sat staring at the Thames smoking weed in my teens on the stone steps that prop up the Ancient Egyptian obelisk Cleopatra's Needle, on Victoria Embankment in London, wondering if my future lay in the buildings lining the streets behind me, or in other countries where the murky meanderings of the river might lead. Later, after I'd first taken the route of the water, then reverted back to those buildings after all, I smoked cigarettes against the imitation bronze sphinxes that flanked the Needle, in lunch breaks from my Whitehall desk job working within the Armed Forces Personnel secretariat, my first Fast Stream posting at the MOD. After work, taxi rides home from stripclubs and nightclubs let me glimpse it as I tried to stare at the apricot-coloured night city sky from steamy windows.

Cleopatra's Needle, poking London in the middle, never to be fully absorbed into its commotion, always
~~~

apart. Despite its phallic bravado and what I knew of its militarised origins and history, its current title made me smile. A woman from the Middle East could turn up in the centre of imperial Britain and take up space as long as she was a Queen, and represented by a granite needle. But still, that was better than nothing, so I thought.

But the obelisk has nothing to do with Cleopatra.

Its naming could be seen as a precursor to our current time's incessant leaning on the PR of 'inclusivity' without ever changing the fundamentals of what it has supposedly diversified. An 'exotic' woman's name next to a possessive noun that cleverly hints at softness, necessity and unexpected feminist power, to describe an enormous monument built to celebrate the self-professed greatness of men.

~~~

In 1819, Muhammed Ali Pasha was the ruler of Egypt and Sudan. He offered Britain the obelisk as a gift to commemorate its victory over France in the Napoleonic Wars, because he was broke and needed favour with those who were not. Maybe the Pasha was calling Britain's bluff, knowing it wouldn't pay for the transportation costs and not offering postage and packing as part of the gift. He was right and wrong. It took almost sixty years for the
~~~

money to be privately raised to transport the obelisk from Alexandria to London. It travelled in an elaborately designed iron cylinder, the monument becoming its own vessel, towed along by a steamship from which it came free during a storm in the Bay of Biscay. Six sailors from the steamship tried to get it back but were drowned under the waves as the obelisk surged away.

Despite weighing ten times more than the heaviest of the stones that comprise Stonehenge, miraculously it didn't sink and was found floating on the waves a few days later by Spanish sailors, and eventually made its way to the Thames and Victoria Embankment.

Despite being half buried in sand for hundreds of years, this stone obelisk had a long relationship with water, from its first journey around 3,500 years ago when it floated on a huge raft down the Nile, from Syenite to Heliopolis, a city of temples where Pythagoras, Moses and Plato studied. Ancient thinkers often accredited with 'Western' philosophies, who studied those philosophies in Africa. Fourteen hundred years later it sailed the Nile again, to Alexandria on the order of Augustus, to adorn the Palace of the Caesars. An obelisk commissioned by a man, Pharaoh Thutmose III, to offer the gods his thanks for being their divine descendant and to humbly ask for a healthy life of thirty years; added to by another man,

Pharaoh Rameses II, who wanted to inscribe his military victories on it. Ordered to be removed to a palace by another man and eventually to London by many more men, some of whom gave it the woman's name 'Cleopatra's Needle', as the floating container that transported the obelisk from Alexandria had been called *Cleopatra*, and perhaps because Alexandria was most famous for being Cleopatra's city.

In 2019, another man, Egypt's most internationally well-known Egyptologist, Zahi Hawass, was furious that Britain was not, in his opinion, treating the ancient monument with the reverence and respect it deserved, saying it should be in a temple or a museum, not forgotten by a riverside. He never explicitly called for it to be returned, but stated publicly that 'if they don't care, they should give it back'.

Of course, that isn't likely to happen anytime soon, despite the 2020 toppling of statues built to celebrate the colonialism that artefacts like the Needle also represent. I'm sympathetic to Hawass's view in adulthood, but I wouldn't have wanted it to go anywhere when I was younger. It was a touchpoint for me, connecting me to the Nile, to the Thames, and back again. A welcome reminder of one of Egypt's most famous women. A woman who had killed herself long before the obelisk would make its way to the palace she had supposedly commissioned to honour Julius Caesar.

~~~

Cleopatra is probably the most often represented Egyptian in English-language storytelling. Her heritage is reflective of the country's then constantly changing leadership, and on her father's side she is thought to have been ethnically Macedonian and Greek, possibly Persian and Syrian, though born in Egypt. Her mother's heritage has never been confirmed. This ethnic ambiguity, as well as a general propensity to whitewashing, has enabled Cleopatra to be played by Elizabeth Taylor, Hildegard Neil, Vivien Leigh and Claudette Colbert. Almost sixty years after Elizabeth Taylor played Cleopatra, Gal Gadot will be the latest non-Egyptian to reprise the role in a Hollywood film to be made by Paramount Pictures.

When I worked as a stripper during my first years at university, every dancer had to choose a stripper name, as is customary in the industry. This particular club encouraged strippers to choose the first name of a celebrity they'd been told they looked like – Kate, Beyoncé, Rachel, Jada. Although the celebrity I've been most told I resemble is Helena Bonham Carter, the manager chuckled when I made this suggestion:
~~~

– Men are gonna go crazy for the Egyptian thing. Use it. Learn belly dance, get some coin bangles or whatever. I'll put you down as Cleopatra.

My British girlhood, when ideas of what is desirable imprint indelibly on the mind, just 'missed out' on the feverish objectification of mixed-heritage and non-white celebrity women that began in the mid-2000s, so I thought he was high. Men did not go crazy for Egyptian girls. But he was the manager and I was a teenager just out of an abusive relationship, desperate for money to survive, pay debts and get educated, so I did what he said. He was right. Perhaps the recent post-9/11 villainisation of all things Arab gave the kind of men (rich, mostly white, mostly American) who went to this 'high-end' London stripclub an extra thrill, paying an obedient, apolitical young Arab girl to take her clothes off for them. Continuing a legacy of colonisation and false economy with a nipple right up in their face. I'm not abdicating my own role in welcoming and encouraging this colonisation and false economy, it was working well for me at the time. It enabled me to claw back a sense of sexual power after years of sexual violence. It gave me a sense of family and belonging. It gave me enough money to study at and graduate from a top London university with zero financial support from anyone, apart from two highly appreciated Hardship Loans from the college. It also gave me more long-term cynicism I could have ever imagined about sex and men. The daily sexual assaults and harassment became impossible, at the time, for me to extricate from my job role and from the eager, fledgling feeling of empowerment. One moment I would be feeling unstoppably queenly as a man with fame and riches showered me with whispered compliments and club tokens (no cash allowed) and then, the next, utterly tragic and panicked as the very same man would suddenly sneer at me, push his hand between my legs and laugh as I jumped back, scrambling for my scant, sequinned coverings whilst calling for the bouncer, who in turn smiled as he accepted his own handful of real cash (no tokens), more than I'd been paid to be assaulted. The cycle of gendered violence, assault and intimidation was one I had grown accustomed to during my teenage years, and though I then saw stripclubs as an escape from this cycle due to the relative sense of empowerment, it was likely just a more financially rewarding version of the skewed normality I'd internalised. These occurrences did become more virulent and my reactions more rageful. I stopped asking the bouncer for help and delivered my own slaps and take-downs, with varying effects. Still, these physical and sexual assaults seemed easier to deal with than the verbal barrage of a 'sit down', which is when a customer would pay for an hour of sitting down with a stripper and often require no stripping, just listening. Sexual fantasies, family or business troubles were the most common subjects, but political lectures were surprisingly not far behind. Though perhaps this was particular to me, 'Cleopatra', who was pointed out to any customer looking for someone 'exotic', 'not European', occasionally explicitly 'Arab'. I never asked the other strippers what they talked about with their regular sit-down clients. That seems bizarre to me now, but at the time, as soon as a shift was over, we all just wanted to count our cash and forget about it. I chose to take £500 for an hour's 'sit down' to listen to pro-Bush speeches from men called Randy, but eventually I would prefer to be writhing naked in Randy's face than listen to him spout the heroics of American interventionism, even if the money was the same. He'd talk about Bush wanting to win a war on terror for the good of the world, that humanity would owe America whatever peace it enjoyed, and I would begin to lean forward, peel off a strap. He would stop talking and sit back. My body as a barrier to politics I disdained.

I have always found it frustrating that until relatively recently, whilst this whitewashed Cleopatra casting could pass without question in the Western mainstream media and I could use it as my stripper USP, outside of the stripclub and film world I would be told by affronted white people on a daily basis:

– I would never have guessed you were Egyptian! Or: No! You don't look Egyptian at all! Or: You must only be half; you look totally English to me.

It wouldn't only be white people who would say this, but they were the vast majority. And it was very much a UK thing. British people of all ethnicities (apart from British Arabs/Middle Easterners and those of mixed heritages) seem to have a far narrower idea of what a Middle Easterner can look like than those in other places I've worked in, especially America and France. This might be because those countries have a greater number of the Middle Eastern diaspora and so they've become more attuned to the huge breadth of aesthetics that come with the region's population.

White people deciding what ethnicity non-white people are 'allowed' to be plays out in all aspects of Western life. You must look 'X' enough – to them – in order to truly present as 'X' and in order to truly know more than them about what it's like to be 'X'. Perhaps you need to speak a certain way, wear certain clothes and have a certain family set-up as well. They will decide if you are allowed to inhabit your ethnicity or not. It may not be fully conscious, but this is a classic survival method for white supremacy. Considering white people are a global minority, for whiteness to remain supreme, they must make as many people as possible believe they are white – but only ever up to a point. Only so they'll feel discomfort at identifying with and ultimately joining those who are challenging racism and white supremacy. Not enough so that they feel entitled and deserving enough to freely pursue whatever it is they want to do, in the way they want to do it – in politics, business, art, education or their home life.

As people with darker skin shades are facing such violent and unrelenting prejudice in all industries and areas of life across the world, it can seem pathetic to try to demand that you, as a white-passing person – and your characters, in my case as a theatre writer – should not be represented as white, nor as any other interchangeable ethnicity. Whiteness, even an association and adjacency to it, protects and privileges. So many Middle Easterners, Arabs and North Africans, especially those who are mixed, can be white-passing; so many cannot. But whilst it is not as urgent as other prejudices, it is not pathetic, and we must be able to inhabit our heritage and demand it be represented, regardless of who finds that difficult.

The British came to the countries where my non-white heritage comes from. Egypt, Guyana. They carved up the land, dug out the gold, cut down the trees, categorised the people into shades and names and ways of living, and let them know that however close they could get to the top (the top being British colonial officer), they would never actually be at the top, for all the blue eyes and half-whiteness they might claim.

They came along
made us the mix we are
now they come along and say
we don't look enough like the mix we are
to really know what it's like to be the mix we are
we don't know who we are
but you do because
'we made you
and we will continue to know you better
than you know yourself
we will continue to tell you better
than you can tell yourself'.

We can't let that continue, but it is difficult to pick apart and know how to stop it. It is internalised to such a degree that I work on projects daily where I cannot be sure if I'm completely comfortable with the way multiracial and

Middle Eastern characters are being represented – and often that is the unsaid reason I have been employed. If I'm not comfortable, how much of that discomfort comes from the way the background elements of a character are being explored – or, more likely, underexplored – and how much from a story that I am just not that interested by, or from the intention to create discomfort in the reader/viewer in order to challenge them. And how can I possibly be responsible for authenticating a piece of work that isn't mine, that conjures up hundreds of fictionalised lives that each have their own backstories and identities, taking place in numerous locations which in turn each have their own myriad histories?

I try to challenge these limiting representations within the industries I work in, mostly TV, film and theatre. Many British and American TV shows, films and theatre productions have been set in the Middle East over the past few decades but they have rarely been made by creatives from the region. I love the work I do, and I love watching these forms of storytelling on a daily basis. But the forms themselves come with an imperial past that is still interwoven into their present. Theatre for example, in the Western way that I create it, was a major tool of imperial propaganda. The British Empire used theatre as a space to rehearse for colonialisation and to recruit the

British public to the endeavour. The late-Victorian era saw a huge number of musical comedies being produced in the UK that were set in India and often made Indian insurgency against the British the lynchpin of the comedic narrative.[10] A staple in the Great Exhibitions of Britain, from the Victorian era until the Second World War, was an immersive theatrical experience called 'Cairo Street'. Belly dancers performed in small theatres along a re-created 'Egyptian street' with shops, a bazaar and camels.

~~~

The timing is no coincidence. The British, with an Anglo-Indian army, first invaded Egypt in 1798 to expel the French, who were attempting to take over Egypt. The British feared this would lead the French to have increasing influence in India, due to the established trade routes through Egypt to Europe. The French surrendered in 1801 and the British occupied Egypt until the Egyptian leader Muhammad Ali defeated them with his Mameluke army in 1807.

Seventy-five years later, in 1882, the British invaded again, this time with the French, and began to colonise the Nile region, ushering in the Nile gunboat era, when Victorian gunboats facilitated Britain's imperial control of the north-eastern corner of Africa, policing and attacking
~~~

from the water, 'floating symbols of British imperial power'.[11]

This 1882 invasion and subsequent military occupation was not about the Nile in the way that the later British invasion of Sudan was, but primarily about the Suez Canal. An engineering feat like no other at the time, designed and managed by the French but built between 1859 and 1869 by the Egyptians, many as forced labourers. Up to 120,000 out of the 1.5 million workers died whilst building it.[12] The Khedive of Egypt had got heavily into debt to fund the canal and other modernising projects, with money borrowed at hefty interest rates from European banks. When the British saw that the canal would be realised and provide a route to India which cut out the thousands of miles and extra weeks to sail around Africa, they bought the now-bankrupt Khedive's shares in it. The French and British began to control all of Egypt's finances, railways, ports, post offices and museums – all due to Egypt's bankruptcy, brought about by the crippling European debts taken on by the Khedive to pay for European projects implemented in Egypt, supposedly for the long-term prosperity for the country but clearly with far more short-term benefits to Europe.[13] The Egyptian army officers were understandably not happy about this set-up and overthrew the Khedive. The British invaded, fearing being blocked from the now-all-important Suez Canal and not trusting the French to support their access, though the French had also sent troops to fight on the British side. After the British Navy bombarded Alexandria from the Mediterranean Sea and attacked both ends of the Suez Canal, Colonel Urabi, who was in charge of the uprising, surrendered and Egypt was now a virtual British colony, allowing Britain to also effectively seize control of Sudan, beginning the European 'Scramble for Africa', which saw 90 per cent of the continent under European colonisation by 1914.

The British had invaded with an Anglo-Indian army and committed more resources to ensure a victory than the French, and so, to the irritation of the French, became the de facto rulers of Egypt, without it becoming an official British protectorate until 1914. For thirty-two years, the British ran Egypt's finances, government personnel and Armed Forces, yet it was ostensibly part of the Ottoman Empire. This fiction suited the British at the time, showing how the confusion and lack of clarity so intrinsic to modern-day international politics has long existed as a means to control the narrative and maintain power.

In 1904, the French and British signed the Entente Cordiale, which formalised Britain as the dominant power in Egypt, whilst France would rule Morocco, Tunisia and Algeria.

Egypt's politicians and population were consistently against the British occupation, and the insidious way they had increased their powers in a country they kept saying they would soon leave. As with all colonised regions, Egyptian nationalist movements and uprisings were a constant threat to British rule. It took hundreds of attempts, the decline of the post-Second World War European powers and the emergence of the exceptionally popular socialist Arab nationalist Gamal Abdel Nasser, who led the Free Officers' military overthrow of the British-backed monarchy in 1952, to finally rid Egypt of British troops in 1956. Though even today, they are regularly stationed in the Sinai and Alexandria to train the Egyptian Armed Forces.[14]

~~~

With constant uprisings across the empire, the British government wanted to employ as many tactics as possible to ensure complete support at home. The Great Exhibitions mentioned earlier, with their immersive, touring theatrical performances of 'Cairo Street' and other 'exotic' wares, were used not only to encourage domestic revering of British rule over otherwise 'uncivilised'
~~~

places, but also to encourage the British population to imagine their role in keeping it going. It isn't an exaggeration to say that theatre and storytelling were ways in which the violent economic realities of empire were translated into adventure and a vital moral mission for the British public.[15] Empire itself is a form of theatre, the superiority of the colonisers always a performance that the performers must believe to keep the show going, so perhaps it isn't surprising that the art form has been used in these ways. It was also used to decolonise, and I hope it will continue to be, at least in its attempts to speculatively and imaginatively recover voices of those colonised by the British, which were purposefully erased or unrecorded.

Ironically, in Nasser's post-colonial Egypt, after he was officially elected president in June 1956, he also used theatre of a Western style to mobilise national sentiment after overthrowing King Farouk and initiating the process to end the British occupation. Although this strategy did assist the nationalist effort to instil pride in an Arab identity after years of colonial inferiority, and many regard it as Egypt's golden age of theatre, it was still accompanied by regular raids and checks on content by the authorities.

~~~
~~~

As a kid, Gamal Abdel Nasser was a name that would always come up within a few hours of landing in Egypt. The conspiracy theories people raised over kunafa; primarily how the British had organised the failed 1954 assassination of Nasser by a Muslim Brotherhood member as he gave a speech celebrating the agreed withdrawal of British troops from Egypt after seventy-four years.

– So obvious ya, Sabrina, listen:

Britain was desperate. India had won independence in 1947, the Second World War had decimated its economy, the empire was in freefall. Egypt provided it with control in the wider area, as well as continued economic power from the Suez Canal. Nasser had managed to harness the vast country's anti-colonial sentiment into cohesive demands that had the support of the entire Arab region. Britain had to try something. This socialist nationalist was succeeding in bringing Arab countries together to fight for their sovereignty over the resources of their land and waters, one of the few things allowing Britain to remain a leading world power despite its crumbling post-war empire. Mostly, he was threatening to renationalise the Suez Canal, arguably the stretch of water most important to the survival of what was left of the British Empire even if the route to India was less of a priority than it had once been.

Of course (the conspiracy theory would go), the British planted a hitman, hid the motivation behind religious ideology and hoped to put a stop to the inevitable loss of Egypt from their shrinking portfolio.

The assassination attempt failed, with all eight shots missing Nasser, even though the gunman was only twenty-five feet away. Depending on who was speaking, this failure was either deliberate – the man finding his loyalty to Egyptian freedom at the last minute – or just a case of terrible recruitment by the British.

Whatever the truth, it was this failed assassination of Nasser which ironically cemented the association of the Muslim Brotherhood and the West in many an Egyptian mind. Ironic because the Muslim Brotherhood came into existence in 1928 in direct opposition to what many believed to be the 'morally corrupting' British imperial rule of Egypt. An ironic association that remains understandably alive today, after the Muslim Brotherhood were backed (and, many would conjecture, put in place) as leaders of an Egyptian government by the Obama administration in 2012, only to be overthrown by mass public protests and the military, led by General Sisi, in 2013.

It was surreal watching the liberal Western world celebrate Trump's 2020 election loss to Biden, and at the same time listening to Egyptian friends lament a Democratic victory, feeling that this would mean an eventual return of Muslim Brotherhood leadership to the country. This is not to say they are supporters of the current regime, but the return of a Western-supported Muslim Brotherhood brings fears of civil war in some circles.

After contraband whisky, the conspiracy chat could turn more TV-worthy, in arguments over the likelihood of curses and mysterious powers having been uncovered by British archaeologists – or robbers, some would say – whilst digging deep into Egypt's deserts:

– No, let me tell you, it's naive to think those Englishmen had subjected themselves to years of sun-scorched scavenging just because they enjoyed the thrill of uncovering something untouched for thousands of years, come on!

– Ha, perhaps you have never met a posh Englishman, then!

– Of course they were looking for something that gave them even more power, the magic of the ancients, ***wallahi***, how else does such a tiny speck of an island still dominate our lives?

– They just like to imagine they own everything, and everyone, that's it.

– They were cursed of course, maybe they found the

secret power, but watch, it will be their downfall, the years will show.

– Yes, you are right. Those pharaohs did not mess around. Another glass?

I loved these dinner-table conversations. I never knew if they spoke about the British as much when I wasn't there. I hoped they didn't, that it was somewhat for my benefit and without me there they would revert to their absurdist, difficult-to-translate Egyptian jokes and gossip about political and cultural figures I wasn't as familiar with as Britain's.

~~~

Back in Britain as an adult, it was working in the MOD which reignited my interest in telling the other sides to the colonial mainstream narratives, the ones I'd listened to at dinner tables and thought about whilst falling asleep to car horns and the buzzing of mosquitoes. An interest that, as I've said, really came into focus during the interviews with the Developed Vetting Officer, my just-doing-his-job interrogator. His extraordinary interest in the Egypt of me, how I related to its water and its people, made me dive deeper than I had before. His tack surprised me at first. Despite its gargantuan status, contesting, at 6,650 kilometres, with the Amazon the
~~~

title of 'longest river in the world' and being the main source of water for the whole of Egypt and Sudan, it was not the Nile which interested him. It was the Suez Canal that seemed to obsess him, and consequently me, post-interrogation. At only 193 kilometres it is a tiny little line really, in terms of major global waterways. It was a name I'd always known, but not paid much heed to, uninterested in its unromantic chains of oblong cargo ships, naive about the role it had played and still played in Egypt's and Britain's pasts and presents, ignorant of how it would suddenly appear as an immovable obstacle in my career trajectory, despite my never having set eyes on it, never having touched a drop of its waters.

~~~

– Your father. Born 1956, Cairo. Quite a coincidence, isn't it?

– In what way?

– Do you mean to say he never sat you on his knee and told you about the Suez Canal Crisis? How he must hate the British for what we did? And France and Israel of course, besmirching his birth year with our invasions –

– Have you met him?

– Who?
~~~

– My dad? I take it you didn't get hold of him as part of this process. Cos if you had, you'd know.

– *What is it I would know?*

– Sir, if my dad sat me on his knee as a kid, it wouldn't be to tell me about the Suez Canal, it'd be to sing karaoke.

But there was no let-up. Sex work and the Suez Canal were the two areas the man in the beige mac was not leaving alone. I could indulge him on the sex work, but I was embarrassed about how little I knew about the Suez Crisis. He seemed to think this was a double bluff. I gave him everything I could to convince him my dad was a man of parties, meditation, music, meandering walks and confrontation avoidance. He had no interest in besmirching Britain, in either actions or words. He loved Britain. Especially London. His popularity with the women more than made up for the racism he endured and he tried to gently instil in me a love of all things British and a rejection of all things Egyptian, including Arabic. He felt my life would be easier and more successful if I could manage to see myself and, most importantly, be seen as 'British British'. After 9/11, he suggested I take my mum's maiden name. Not a 'British British' name either but one which revealed a Latin American identity, something which my dad thought would serve me much better in this Islamophobic anti-Arab future we were seemingly crashing towards. My feminist brain enjoyed the thought of taking my mum's name and I had to admit it had a ring to it – Sabrina DaSilva. Whilst it is undoubtedly a luxury to be able to consider choosing between names that you feel you can own equally, Mahfouz was my name, it was what I had grown up with, it was who I had been seen as and I couldn't name myself out of those experiences, even if I wanted to. As 9/11 made my name and my heritage suddenly visible in a way I'd never anticipated, it was up to me to make sure I did not become someone who shapeshifted to suit white supremacy and all its ugly violence. I failed in this endeavour in many ways, but I did keep my name. My dad nodded his understanding.

I didn't tell the interrogator this. I thought he might see my love and defence of my Arab Muslim surname as evidence of my extremism. I also wanted to avoid a similar reaction to the DaSilva name. Guyanese made up of Amerindian, Madeiran, West African and French heritage. Guyana, a South American colony to which Britain had sent troops in 1953 to suspend the constitution and bring down Communist rule. My grandad was a teenager there during those times and talked fondly of

Marxist–Leninist policies, whilst reserving his most passionate oratories for Fidel Castro, the most mentioned political figure of my childhood and one to whom I was instinctively sympathetic, until I went to Cuba myself and met few people who shared my grandad's glowing views. Strangely, none of this part of my identity came up. The interrogator had all my family names and histories in the file. It was a sign of the times that 'the threat of Communism' had taken a back seat to 'the threat of Islamism' and was treated as such when questioning someone with potential familial ties to both. I realised that whatever happened with my career at the MOD, I needed to know much more than I currently did about Britain and Egypt's intertwined history, especially now that I was seen firmly on the side of 'other', and I'd better begin with the Suez Canal.

~~~

Nasser renationalised the Suez Canal in 1956, declaring it Egyptian-owned and -managed. The Suez Canal Crisis ensued. The British, French and Israelis made a secret deal, the Protocol of Sèvres, agreeing that Israel would invade the Sinai on 29 October 1956, and following that Britain and France would mobilise as 'interventionists'. When they did, they began what was arguably one of
~~~

the most significant episodes in post-1945 British and Egyptian history. Though the tripartite coalition were successful in their military endeavours, with Israel occupying the Sinai until March 1957, it was a political failure, as all three were forced to withdraw from the Suez by American and Russian pressure.

A few days after the invasion, on 4 November 1956, the United Nations threatened Britain with sanctions if there were any civilian casualties from British aerial bombing of targets in Egypt. This led to economic panic in the first week of November 1956 and resulted in tens of millions of pounds being lost from the country's reserves. Britain faced having to devalue its currency. Meanwhile, furious that military operations had begun without his knowledge and against his wishes, especially considering the supposed 'Special Relationship' between the UK and US, President Eisenhower made sure the International Monetary Fund denied Britain any financial assistance and sold the US government's pound sterling bonds.[16] Russia had made it clear that Western powers had no right to interfere with the canal's nationalisation, unless Nasser refused to pay the company's shareholders or blocked navigation of the waterway. Nasser had done neither, something which is less documented in the British accounts of the crisis, even the highly critical ones.

As the situation progressed, the USSR made it clear that it would not 'stand aside',[17] giving its reason as concern that instability in the area would affect the stability of the USSR. Eisenhower argued that the Russians were actually getting involved in order to divert international attention away from the concurrent uprising in Hungary, which they were brutally supressing. Though it would deny this, the growing unrest across the USSR meant that the Soviet government was not in the mood to lose any of its allies. As Nasser was considered to be its main friend in the Arab world, and as nationalisation was also a key ideology in its Cold War positioning, the Soviet government would have to do what it could to stop Egypt's president being toppled. The First Secretary, Nikita Khrushchev, declared that the USSR was ready to send troops to Egypt and to launch nuclear attacks on both London and Paris if British and French troops were not withdrawn. Regardless of whether this was posturing or not, popular opinion across the world, and particularly in the Commonwealth countries, was horrified at the coalition's invasion. Only Australia supported Britain, whilst Pakistan threatened to leave the Commonwealth. In Britain itself the conflict divided opinion. The Conservative government faced significant hostility from the

Labour opposition and experienced a split in its own party. Nationwide anti-war protests were organised and several civil servants resigned in protest.

Completely isolated and humiliated, the French President, Guy Mollet, and the British Prime Minister, Anthony Eden, reluctantly accepted a UN-proposed ceasefire. Under Resolution 1001, on 7 November 1956 the United Nations deployed an emergency force (UNEF) of peacekeepers into Egypt to halt the conflict. It had lasted just two days, and yet it would often be credited as the crisis which ensured the end of the British Empire.

Egypt maintained control of the Suez Canal with the support of the United Nations and the United States. The canal was closed to traffic for five months due to ships sunk by the Egyptians during the operations. British access to fuel and oil became limited, which resulted in shortages. Petrol-rationing in the UK was introduced in December 1956, lasting until May 1957. Under huge domestic pressure and suffering ill health, Eden resigned in January 1957, less than two years after becoming Prime Minister.

The Suez Crisis understandably increased Soviet influence over Egypt. Khrushchev's intervention on the side of Egypt and its public proclamations in defence of its sovereignty against colonisation placed the Soviet Union as the natural ally of Arab nations. This Soviet support fortified Arab nationalists, and Nasser began to aid rebel groups hoping to gain independence from the British across the Middle East.[18]

The crisis wasn't just a historical event. It was both the real and the metaphorical fatal blow to Britain's centuries of global dominance. And as I'd find out in that stuffy little Whitehall room over fifty years later, it would not easily be forgotten.

~~~

From 2015 to January 2020, the Foreign and Commonwealth Office posted the following on its Egypt page:

> The Suez Canal Zone (SCZone) project is estimated at £20 billion over 15 years. The project presents UK companies with significant opportunities in:
>
> - ports and logistics development
> - development of a 76,000 sq km industrial and logistics hub
> - new industrial zones and urban areas
> - new transport infrastructure
> - power generation
~~~

- water desalination
- waste water treatment plants

Other opportunities in the construction and infrastructure sector include:

- power infrastructure – Egypt plans to invest USD 110 billion up to 2027
- waste water plant expansions
- water infrastructure PPPs projects
- tourism infrastructure – in April 2014, the government earmarked USD 136 million for the tourism infrastructure
- airport city at Cairo International Airport with an investment of USD 20 billion
- New Cairo Capital city – a £30 billion project[19]

The above opportunities were removed from the site in January 2020 and the page's current guidance on British trade in Egypt gives no details on any of these areas. However, in June 2021 the UK Prime Minister's Trade Envoy to Egypt, Sir Jeffrey Donaldson MP, detailed the UK's commitment to investing in the Suez Canal Zone as an integral part of the UK's objective to be the largest G7 investor in Africa by 2022.[20] Clearly the waters of the Suez are as important to Britain as they ever were.

Harbour I (Egypt)

Though I had never stared at its surface, and it was actually the toxic froth of the Thames in my every connective thought, I began to understand that this would never be enough to be truly, reliably British.

***

Dunkirk salt spray seals my cavities
rotten roots a botanical legacy
of Somme River weeds.
Rote learn those ruptured dates, GCSE palm sweats.
Politics seminar role play, David Lloyd George.
Autumn pound coins for poppies.

From my Whitehall office window,
Cenotaph inscribes sacrifices wreathed every day,
I cry proudly for their creases.
The man asks me;

Are you not made of Suez silt?
How do we know you won't shore our boats
by making yourself bigger than we made you?
Thames. Suez. The Nile, too.
What greedy hydrology you have?

Actually, we shouldn't conflate rivers with canals
I say, but this can't help my cause.
British intelligence always tells water what to do.

Sun slants in, I dip my biscuit;
Look, sir, I want to make the world a better place,
peace-keep with the nuance of my name.
His smirk between beige mac lapels:
But we do not work for the world, my dear.

***

U.A.E

SAUDI ARABIA

OMAN

YEMEN

Sanaa

Aden

Arabian Sea

Socotra

SOMALIA

Tidal

(Yemen)

Have you ever had sex with an animal?

~~~

He asked me bluntly, his pen poised to tick or cross the box. His face rearranging as he said it so that he suddenly appeared tired, perhaps of this question, perhaps of the answers he had been given by others over the years.

At this point in my life, I am used to sexually loaded questions, innuendoes, outright demands or requests. It was only a year since I'd stopped working in stripclubs, and after five years of being face to face with countless older men who didn't even blink whilst blurting out their sexual desires, I imagined myself immune to the type of freeze that was coming over me in that small, Whitehall meeting room.

He leant back, smiling his tight smile, enjoying the anticipation of the duty that would now fall to him, but giving me a little more time to say something, anything. I did not.
~~~

– It is a standard question, Ms Mahfouz. One which, I'm aware, must seem rather strange. But you must understand that this is not a process to assess whether you are suitable for the role. You have already got the job, done the training. This is simply to assess whether you would be liable to bribery and therefore treason. Bestiality is far more common than you might imagine and, understandably, something that could be used as leverage against someone. An affirmative answer to anything I ask you is not a judgement nor will it affect your employment in any way, as I have already said. But we must know, so that we are aware of, and can mitigate against, the risks.

– No, sir, I have never had sex with an animal.

And it is a lie. Because so many of the men I have had sex with are animals, in their own ways. Some were straight-out rapists. With others, sex that was seemingly consensual at the time, through retrospection became an emotionally coercive demand on my body at best.

Like many people who have experienced sexual violence, I am often rendered incapable of anything but guttural emotional responses when I hear of it happening to others. Ten years after being asked if I had ever had sex with an animal, whilst sitting in the audience of a panel discussion regarding the conflict in Yemen I am reduced to a blubbering wreck when I hear an aid worker estimate that 98 per cent of women in Yemen have been sexually assaulted during the current conflict. There are no official figures because even if women have access to means of reporting an assault, it is highly discouraged and is potentially life-threatening. The estimate was made by taking into account statistics from the 2013 Demographic Health Survey[21] in which 93 per cent of women reported violence against them in the home as commonplace, and this was before the conflict began. With a 2019 UN report declaring the situation in Yemen the largest man-made humanitarian crisis in all of history,[22] it seems likely that all injustices faced by women will have increased exponentially. The UN report also states that 22 million people require aid and that 100,000 children had died of starvation over the four years. The numbers are viscerally staggering. Replace 'Yemen' with the name of any country in the global north and the evidence of human value being politically and economically determined has never been starker.

In 2018, a year before this UN report was released, a bomb manufactured in America by Lockheed Martin, a British American arms company, that had been sold to and used by Saudi Arabia under an arms deal struck in

2015, killed forty Yemeni boys aged 6–11 who were on a bus coming home from a school trip.[23]

Saudi Arabia is the single biggest customer for both the US and the UK arms industries, worth around £6.4 billion to the UK in the last five years.[24] By 2020, the British were still sending arms to Saudi Arabia, despite a Court of Appeal ruling that Liam Fox, Boris Johnson and Jeremy Hunt acted unlawfully in 2015 by secretly signing off on arms sales to Saudi without properly evaluating the impact on civilians.[25] The court ruled that it was illegal for Britain to sell any more arms to Saudi whilst there was a chance they'd be used against civilians. It is clear that the supply of arms from the UK is integral to the Saudi offensive and therefore the immense humanitarian crisis in and destruction of Yemen, yet shockingly more than half of British people are completely unaware of the conflict, let alone the British role in it,[26] which is even more saddening considering that this British exploitation of Yemen started way before the illicit arms deals of the twenty-first century.

In 1839, British forces invaded Aden, at the tip of what is now Yemen, and forcibly took control of the area, as it was perfectly placed for them to launch attacks on the pirate ships in the Arabian Sea which had long been attacking the passing British cargo ships. Importing

British goods to India and exporting Indian goods to Britain was, at that time, what made the empire viable. Pirate attacks were a very real threat to the profitability and therefore the survival of the British Empire, so establishing strategic points to thwart these attacks was a top priority. As a consequence of this military presence in Aden, it became a convenient stop for British ships to refuel, and after the Suez Canal in Egypt opened in 1869, Aden became one of the most important ports in the world to the British. It remained so until 1965, by which point it was well known as 'The Last Post' of Empire.

~~~

The BBC made a six-part TV drama called *The Last Post* in 2017, following a unit of Royal Military Police officers and their families in the early 1960s as they deal with a tide of anti-imperial politics, but mostly extramarital affairs. It was well written and acted, based on the writer and creator Peter Moffat's memories of living in Aden as a child in such a unit. However, it does what British storytelling about the Middle East always does: centre the British. Which wouldn't necessarily be a problem, if 'British' was allowed to be broader than a particular type of white British. Of course, all kinds of stories deserve to be told and those colonial officers should be written
~~~

as real people, with conflicted morals and emotions. But their stories have been told so many times that they have become the default understanding of empire. Creative storytelling in books, films, theatre and TV greatly influences a nation's imagining of its identity and history. Even though I grew up with access to both sides of the story, I was besotted with English-language films, TV and books. Despite all other information to the contrary, from consuming these I can easily understand why many in Britain have no trouble imagining a complicated colonial officer, with a full life outside of his professional role, yet struggle to imagine the people fighting against colonial rule as being much more than just that – against colonial rule. These fighters may sometimes be likeable characters that are created to represent an opposition, so there appears to be perspective, to be value given to that opposition. But without those characters being centered as often and with as much space given to their intimate relationships, as much screen time given to their daily mundanities, as much emotional development given to their trajectories, ultimately they just become sympathetic political symbols set against the fully humanised white British colonisers who were just doing their best in an unfortunate situation.

It is the worst kind of criticism to wish a show was an entirely different one from what it was, rather than to assess it on its own terms. So instead of asking what *The Last Post*, which I enjoyed, could have been, I wonder what other shows could have been made and programmed alongside it; such as a drama about Khadija al-Hawshabi, a woman whose family once ruled Radfan, a mountainous area north of Aden. Khadija was part of the Arab Women's Club in Aden, a radical offshoot of a group the British had formed in 1951 called the Aden Women's Group, which still existed but was aligned with colonial objectives. In 1963, an uprising beginning in Radfan marked the start of the revolution in South Yemen, which played a vital role in the 1967 declaration of Aden's independence from Britain.[27]

Khadija became a famous fighter, and whilst not much is documented about her, local stories abound of her decapitating a British captain and displaying his head in the city as punishment for the British not listening to the demands of her family, which consisted of freedom from colonial rule, colonial taxation and the emancipation of their women, whose roles had been narrowed since the British occupation due to the increase in poverty and focus on capitalistic productivity. Khadija was eventually killed in battle as the British sent reinforcements to quell the uprising. However, her violent and revolutionary legacy lived on and the Southern Yemenis gained independence, in 1969 declaring themselves a Marxist–Leninist one-party state, supported by Cuba, East Germany and the Soviet Union, as they aimed to eradicate inequality and colonial influence on their culture, whatever the price. However, British investment in the region had been strictly limited to the small capital of Aden, which was the only location strategically important to Britain. Due to their investment in Aden, the British ensured the resources of South Yemen (North Yemen was under Ottoman rule until 1918, when it became independent) were directed to the capital, including water supplies and infrastructure. For example, 127 paved roads were completed during colonial rule in South Yemen, and by the time of independence only fourteen of these were outside of Aden.[28] The British also created a water supply system that brought water from reservoirs outside of Aden to the capital, and this same infrastructure is still in use today. It is outdated and does not service the contemporary needs of the area, and any plans to modernise it became impossible once the conflict began in 2015. Since then, the system has sustained damage to the value of at least $59 million, as water supplies are strategically targeted, and households are only receiving 3–4 hours of piped water once every few days.[29]

During British rule, this lack of development outside

of Aden, and the siphoning of resources such as water to the city, increased poverty and made life more difficult for people living regionally. This disparity contributed to the many internal conflicts post-independence, with border clashes between North and South Yemen happening from 1972 until unification in 1990 and, from then, continuous sectarian conflict between Shia and Sunni groups, amongst other political grievances. The different sides are supported by richer nations who, like the British, recognise Yemen's strategic location as a guardian of access to the Red Sea and therefore the Suez Canal. Shia dominance in Yemen is also seen as a threat to surrounding Sunni states, as Iran is the regional Shia power. Saudia Arabia, the regional Sunni power, first got involved in 2015, along with Egypt, Morocco, Jordan, Sudan, UAE, Kuwait, Qatar, Bahrain and an American private military company, Academi, in an attempt to defeat the Shia (Houthi) rebels. This Saudi-led coalition continues today, with the UK, US and France involved primarily through the sales of weapons and provision of logistical and intelligence support.

Whilst there are numerous complex and historical reasons for the continuation of Yemen's civil war, access to the Suez remains one of them. It is still the shortest sea route between East and West, and as 80 per cent of the world's traded goods continue to be transported by sea, its vital economic importance remains. Keeping the area accessible and controlled is one of the reasons so many international players have become militarily involved in Yemen over the last few years. If Iran were to dominate Yemen, then it could control access to the Suez Canal, and thereby to one of the world's major trade routes.[30] Unfortunately, the humanitarian response has not been as enthusiastic as the military one, with the UK continuing to illegally sell weapons to Yemen and simultaneously cutting its aid package by 46 per cent, to £87 million for 2021/22.[31] The Minister for the Middle East at the time of this decision, James Cleverly, said to the House of Commons:

> The UK will provide at least – I repeat, at least – £87 million in aid to Yemen over the course of financial year 2021–22. . . . [It] will feed an additional 240,000 of the most vulnerable Yemenis every month, support 400 health clinics and provide clean water for 1.6 million people. We will also provide one-off cash support to 1.5 million of Yemen's poorest households to help them buy food and basic supplies. Alongside the money . . . we continue to play a leading diplomatic role in support of the UN's efforts to end the conflict.[32]

And so we come back to the image of the conflicted colonial officer, doing the best he can in an unfortunate situation.

~~~

I began to see my interviewer as one of these colonial officers from the films and programmes I'd watched and the books I'd read. I interpreted his flushed cheeks for embarrassment over the questions he was forced to ask me by his superiors. I imagined his tight smiles were only tight because his young child was waking him up with nightmares every night and, after he'd soothed them, he couldn't get himself back to sleep again. I wondered about his beige mac, if he knew it was a cliché, to be a white man in a suit doing what he did with a briefcase and a beige mac, the collars turned up, but the belt never used. Did he wear that mac when he went to the cinema?

As he noted down my answer to his question about animal sex, he took his time. Certainly, he did much more than just mark an X in a box.

Maybe he was wondering about me in the same way I was wondering about him. Doodling the shape of my earrings and hoping for a time relaxed enough to ask where I'd got them, his wife would love a pair. Or perhaps it really was all about business. Perhaps he was
~~~

writing down the way my eyelids twitched when he said the word 'sex'.

Perhaps he recorded the many different inflections he'd used to say the word 'sex' since we'd first met all those weeks ago. How dedicated he had been to ensuring the full spectrum of sex had been explored with this woman half his age, who he had warmed up by quizzing for hours about working in a stripclub. Wanting to know every detail – the outfit she wore when she auditioned? The feel of the pole against her skin, did it stick? Why the manager hired her, what he said was special enough about her to work at his stripclub, an exclusive stripclub? Details he needed for the report, of course. And how many of the men she had slept with – staff members, managers, clients?

This was the way sex entered the room. With a refusal to accept the answer that I hadn't slept with any of the staff, the managers or the clients. His little smile, the crossing of his pressed trousers.

Repeating the assertion again. That was surely the way stripclubs worked?

– No, but so what if it were, sir?
– *We just need to know.*

After I'd banged my head on the table, exhausted

with denial and ready to just say yes to anything and everything so I could leave the stuffy room and the crumbs of shortbread surrounding me, he changed tack.

– *That apartment your Iranian 'friend' rented to you. Curzon Street, Mayfair. MI5 HQ used to be there, opposite your building. Saudi Embassy down the road.*

– Yes?

I did not know if I was meant to have been selling sex or a double agent, or both. Part of me wanted to satisfy his ache. Part of me wanted to be the likeable antagonist, the sympathetic political symbol in this episode, rather than a complicated, morally and emotionally conflicted human. I resented my fullness and coveted the interrogations my white male friends I'd met in training had told me about. The smirks over numbers of sexual partners, the immediate crossing off of the 'sex with animals' sections with 'nobody needs to know about that, do they' laughs.

I understood that many of my experiences were not the usual ones of the people they were used to employing. Experiences that stemmed directly from who I am, and they were not used to employing people like me.

– *You have to understand, Ms Mahfouz, it's unusual, the life you've lived.*

After five weeks of protesting at assertions made about me, explaining the basics of Islam, the basics of stripclubs, the complexities of caring for a family as a working-class woman with a postgraduate education, the dot-to-dot of how debt appears in a life of someone who does not have any financial input from anywhere other than her own time and energy, after giving over every one of my vulnerabilities to be ballpoint-validated by a man and an institution who would never, ever understand me, I did understand, finally. And once I did, there was no going back.

***

This is an old, old war
This is an old, old war
This is an old, old war

Blame us for the drought,
blame us for the thunder,
blame us for the waves,
taking us all under.

Burn us for your palace,
burn us for your crown,
burn us for our bodies,
taking us all down.

Tidal (Yemen)

This is an old, old war
This is an old, old war
This is an old, old war

We carry history on our shoulders,
we carry history in our hearts.
We've carried ourselves over mountains,
to end up drowning in the sand.

This is an old, old war
This is an old, old war
This is an old, old war

Blame us for the drought,
blame us for the thunder,
blame us for the waves,
taking us all under.

Burn us for your palace,
burn us for your crown,
burn us for our bodies,
taking us all down.

This is an old, old war.

***

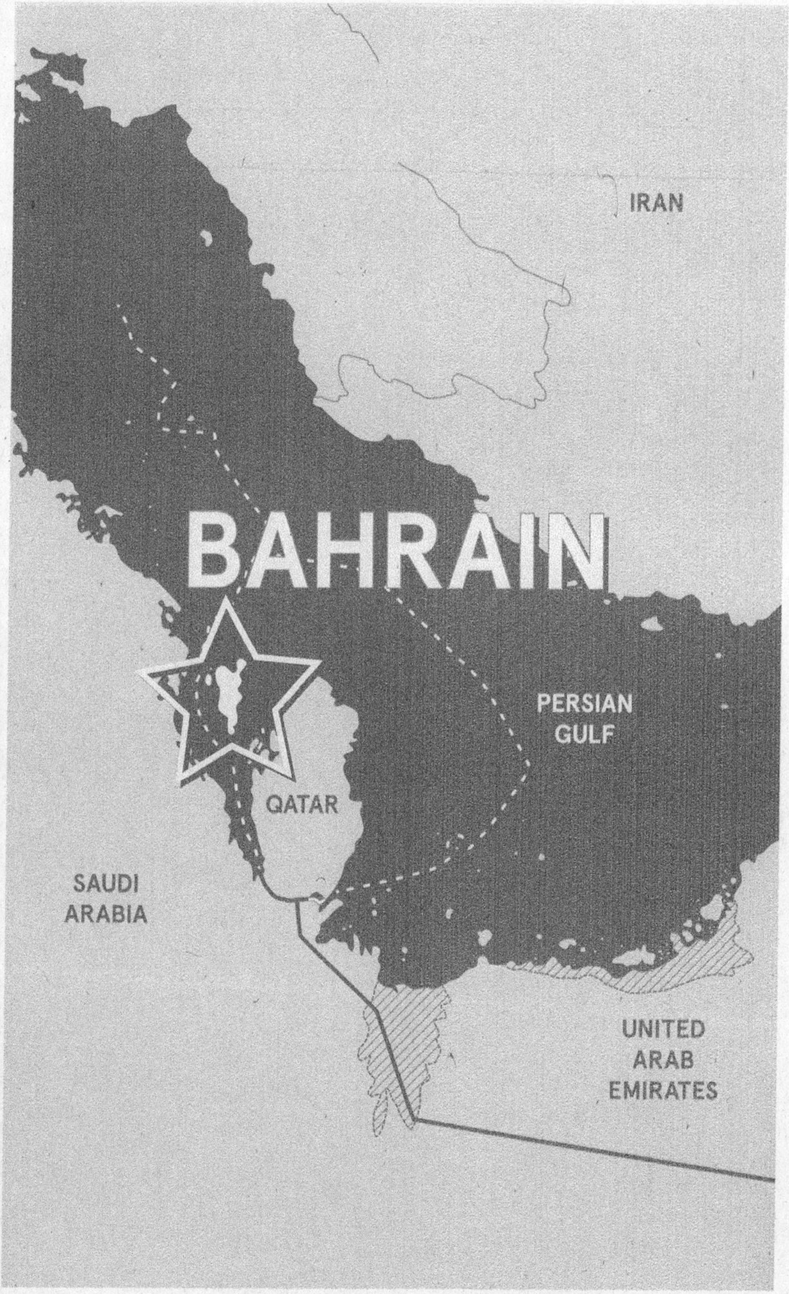

IRAN

BAHRAIN

PERSIAN GULF

QATAR

SAUDI ARABIA

UNITED ARAB EMIRATES

Headwaters

(Bahrain)

Ms Mahfouz, can you admit the British Empire didn't start all of the difficulties in the Middle East? You seem quite hell-bent on starting a region's history rather late in the chronology of humanity, I have to say.

~~~

The vetting officer seemed bemused by my insistence that I could both administratively support the current foreign policy of the UK without politically and philosophically supporting it, and even more surprised that I could do that whilst maintaining conviction that a significant proportion of contemporary problems could easily be traced back to the British Empire. From a purist viewpoint, he was absolutely right. But in turn I was bemused at why and how he'd imagine being critical of certain policies and decisions would make me, as a civil servant, unable to do my job. I'd very much committed to the idea of the Civil Service, despite not being a monarchist. Fundamentally, the Civil Service is a permanent bureaucracy of the Crown, accountable to the reigning monarch,
~~~

although in practice it is in service to the government of the day.

I'd accepted I would have to work under leaders I hadn't voted for and vehemently disagreed with, knowing that whilst I could attempt to push for policies that I found more appealing and inclusive, the decision would ultimately be with an MP who had perhaps never heard of the issue we had been working on for years. I wanted to be one of the cogs that helped the country turn, and if I could make some changes along the way then even better, but I had no grand designs to be an Undersecretary, like many of the Fast Streamers did, or the head of a think tank. The Civil Service appealed in certain ways to the epic and the ego I mentioned earlier, but it was also, in my head at least, intrinsically tied with the word 'service', which I had thought of as a service to the public. A service to people who relied on 'the incorruptible spinal column of England'[33] to fight for their best interests, whatever the dubious whims of changing governments. And I did find that there, a commitment to challenging incumbent governments on policies that did not align with what years of research and expertise suggested would be the fairest, most sensible way forward. When a defence minister put forward proposals on provisions for retired Armed Forces personnel,

I'd attended meetings where civil servants had been the ones highlighting the disproportionate homelessness ex-service people suffered, as well as a lack of satisfactory mental health provision, offering solutions on how to provide housing and health services within the budgets likely to be available. Whether that suited the political agenda of the time or not was beyond their control, but they felt they'd done what they could to fight for the best interests of those that particular department represented. I also found that internal ideas about 'best interests' differed wildly and made successfully challenging the government difficult without a unified stand. But they still did it. It's a process which recent governments have wished to dismantle, with plans for politically appointed heads of departments and an 'office of the Prime Minister', to exert greater control over Whitehall and to make those challenges from the Civil Service far less likely. As political opinion becomes more starkly divided, it is easy to see how the professional political neutrality of civil servants becomes a tougher demand. In 2019, Nigel Farage went on an anti-Civil Service offensive, accusing it of derailing the Brexit process and calling for all Remain-aligned civil servants to be fired if they were seen to be obstructing a Leave agenda, justifying this stance by saying civil servants are there to 'do what an elected government tells them to do'.[34]

Part of what drew me to the Civil Service was the sense that at its best, although government ministers do tell civil servants what they want them to do, they would not just immediately 'do what they were told', but rather would respond with information and research to let the ministers know how viable and beneficial to the public these plans were, in a way that was not only as politically neutral as possible, but also financially neutral, something that private think tanks and consulting companies can never claim to be, given the extortionate amounts most of them charge.

This may all seem overly idealistic more than a decade later, and even then I saw elements of the Civil Service in need of reform, not least in terms of recruitment but also in levels of subject expertise, which often wasn't taken into consideration for senior posts – arguably much like the ministerial positions of government. The recent merging of the Foreign Office and the Department for International Development is an indicator of the structural changes to come. The long-term results remain to be seen, but to relinquish the relative power of the Civil Service to advise on and, if necessary, challenge governmental decisions rather than just blindly implement them would be a huge shame and a crucial step into an even more unbalanced concentration of power in the country.

~~~

In the ironically interconnected way diasporic lives must go, with positive opportunities and benefits never being far from the very thing that caused displacement and difficulties, the concept of the Civil Service, the organisation I admired and worked hard to join, was originally created in response to the expansion of the British Empire and influenced by imperial officials who recommended the Chinese system of civil service that they had seen when posted there. The University of London's School of Oriental and African Studies (SOAS) is where I studied for my postgraduate degree and was widely considered to be the most politically aligned higher-education choice at the time for left-wing-leaning students from the post-colonial diaspora wishing to learn a less Britain-centered curriculum. It was originally founded as the School of Oriental Studies in 1916 to train British colonial civil servants. We are all part of the history that has brought us to the places we occupy and settle into, however welcome or not we may be.

~~~

The Persian Gulf was the body of water that made Bahrain the first place in the Middle East the British really settled into.

By 1858 official British Crown rule, as opposed to East India Company control, was established in India. The Persian Gulf had long been the point at the top of the rich Indian Ocean trade triangle – between west India, the Gulf and East Africa. For Britain to successfully dominate and control trade from India to the Gulf and Africa, it needed to keep control of this area. This was not new. The Sumerians, Assyrians, Babylonians, Persians, Arabs and Portuguese had all asserted control over this body of water for economic and imperial reasons over thousands of years.

By the mid-ninteenth century, Britain had competition. France was also building its empire, and Napoleon's expedition to Egypt, coupled with the Gulf area's 'pirates' (defensive Arab fleets attempting to keep out the British) made Britain take direct action to secure this triangular trade route with the perfectly positioned island of Bahrain. It had informal protectorate status at first, making it one of the earliest countries to become a British protectorate, the first being the Mosquito Coast (in modern-day Nicaragua and Honduras) in 1655. As the name suggests, it was basically a protection racket.

The country would give over control of all its foreign policy to Britain, but without the leaders losing face and becoming an outright colony, and in return British forces would spill the blood of anyone who tried to spill theirs, supposedly. The British were soon aware of how successful protectorates could be for business and by the mid-1800s had forced protectorate status on the Falkland Islands, Hawaii, Nepal, Oman, Penang and Singapore. Formal treaties were signed with Bahrain in 1880 and 1892 to solidify this agreement and, after the First World War, Britain moved its Gulf naval base to Bahrain and began directly implementing reforms across areas ranging far from foreign policy, such as schools, courts and municipalities. These reforms were not welcomed by the ruler and the British replaced him with his son, who promised to support them, and so the 'warm relationship' between the two countries began,[35] continuing through independence in 1971 to the present day, as Britain remains in the country in more ways than one. It re-established its Bahraini naval base in 2018 at a cost of £40 million. When he cut the reopening ribbon, the British Defence Secretary proudly said that the Armed Forces which would be stationed there were 'the face of Global Britain' and by being there, in Bahrain, would 'protect *our* way of life'. What does that actually mean?

That it is positioned between Saudi Arabia and Iran, so it is essential for global energy security? The Gulf shipping lanes continue to need Bahrain's largest body of water to transport 17 million barrels of oil per day. And what exactly is 'Global Britain'? If the face of it is the Armed Forces which are stationed in old colonial bases in past colonial outposts, as the Defence Secretary said, then we need search no further for the body.

~~~

On Tuesday, 20 November 2018, the Bahrain naval base was discussed in the House of Commons. The Conservative MP Leo Docherty, an ex-soldier who wrote a book about his tours in Afghanistan and Iraq called *Desert of Death*, made this statement to the House about the base:

> The base is a huge step forward. It is the first new naval base in the Middle East since 1971, and we should all wholeheartedly welcome it. The base is hugely important to our bilateral relationship with the Kingdom of Bahrain. Many of us know that we have a long-standing relationship of at least two centuries with the kingdom. Because of the pressures it faces due to its location vis-à-vis Iran, the Bahraini state feels a sense of existential insecurity. It therefore relies
~~~

> on its allies to stand with it through thick and thin, and I am proud that this country has done that. Our tangible, permanent commitment to having a Royal Navy presence in the kingdom is of huge importance to our Bahraini friends. In fact, it is so important that they have been prepared to pay most of the costs of the base. That is of huge advantage to us — it allows us merely to man the facility. The reassurance the base provides our ally should not be understated.[36]

The 'protectorate' language and context here are clear. The continuing belief in British superiority was made explicit by a follow-up comment on the part of a Conservative MP, Kevin Foster:

> I agree with what my honourable Friend is saying. Does he agree that we must remember that if we, as a western democracy, do not engage, others will be only too happy to fill the void, as we see with the Chinese military base in Djibouti?

Only Carol Monaghan, an MP for the SNP, even slightly challenged this moral self-satisfaction, by mentioning Bahrain's 'worrying' human rights record and asking that the authorities should be 'pressed' on this. The reply was that she should go there herself to 'see the reality'.

Another Conservative MP, Andrew Bowie, followed up quickly in support of the naval base:

> . . . would he agree it is also a physical embodiment of what we are all talking about, namely global Britain? As we leave the European Union, such things demonstrate that we are not retreating from the global stage, and they are a demonstration of our intent not just east of Suez but around the world.

'Global Britain', 'east of Suez', 'global stage'. Reading these comments by present-day MPs can give the impression that the empire's domination in the Middle East was just a dress rehearsal for what the current British government has planned for the future. The Minister for the Armed Forces, who at that time, in April 2018, was Mark Lancaster, went as far as to say that the base (italics my own):

> . . . epitomises the importance that the UK places on its relationship with Bahrain and the security of the Gulf region, and *the emphasis that the Ministry of Defence is placing on global Britain*. It is the first permanent overseas Royal Navy establishment operating east of Suez in almost half a century, and part of the commitment to the Gulf region that the

> Prime Minister promised in Manama in December 2016, when she underlined her undertaking that: 'Gulf security is our security'. . . the then Foreign Secretary announced that the UK would be spending £3 billion on defence commitments in the region over the next 10 years. It is clear that we cannot afford not to do so — as has been said, 40 per cent of global oil production is shipped through the strait of Hormuz between our close ally Oman on one side and Iran, which is a challenge, on the other. *It is the world's most important maritime choke point.* The wider Gulf contains two more of the world's eight recognised maritime choke points, with the Bab-el-Mandeb at risk of miscalculation emanating from the persistent and tragic conflict in Yemen.[37]

The Middle East's bodies of water are clearly still playing a vital role in keeping Britain 'global', perhaps even more vital now than when it all began, as we have seen with its current involvement in Bahrain's part of the Persian Gulf.

An archipelago in-between two seas. That's what the name 'Bahrain' means – two seas that meet and beef. It was part of Sumeria, one of the very first civilisations in the world. The Sumerian language is the oldest-known written language, dating back at least 5,500 years.

Sumerian myths can be seen as a starter pack for mythology, and one of my favourites, Ninhursag and Enki, takes place in what is now Bahrain.

***

Ninhursag was the Sumerian goddess of mountains, of giving life, despite rocky ground and storms of sand. Her husband, Enki, was the Sumerian god of fresh water, but also the confuser of languages, the god of trickery and intelligence.

Ninhursag needed no evidence to know what she knew; Enki was a god in the way storytellers created them all to be – ego-exploding, freeloading, pyromaniac serial rapists, basically.

Now, Ninhursag wasn't exactly a moral compass – she wanted to amass some wealth for herself rather than get justice for the ones he attacked. Even when Enki began to assault some of the many children they'd had together, instead of damning him from the immortal realm she demanded a city in return for forgiveness. So, Enki guilt-built the city of Dilmun for her, the first recorded story of a city.

But Ninhursag was still not happy: the city was dry, there was no water. If she couldn't protect her daughters she at least wanted to know what else she could grow.

Enki eventually hooked her up with the one good thing he had, fresh waterfalls frothing into rising rivers, ports for ships to dock and load at, and lo and behold this was the ticket. Ninhursag could finally see it. Water, specifically the docks and ports for trading ships, would make her rich enough to rule alone. And so, just like that, the original capitalist imperialist dream was born, with a slightly more feminist undertone than I'm sure anybody thought there would be.

But, as we have all seen, historically and presently, imperialist capitalist dreams don't generally end well.

Enki couldn't stand to see Ninhursag's power grow and so he ate all the plants and trees that she'd seeded in Dilmun. This was the final straw for the goddess of life: she started to kill Enki slowly and painfully. He was begging her to stop, let him live, and finally, for the sake of their history, she did.

But as a couple they were finished. She left the riches of Dilmun, relocating to the searing solidity of the desert, as far away from water and Enki as she could be, and there she planted the Tree of Life. Without any water, it grew.

Roots reaching like flashes of underground lightning, pushing up a gnarled trunk fighting its way to infamy, the only living thing in the vicinity, the only known tree to have grown full formed without any form of water.

Dilmun had taught Ninhursag the ways of water, the power its buoyancy gave her, but as she was a goddess, she herself could now embody it, she could spread across the arid desert with or without it. The Tree of Life still grows today. In the driest spot of desert. Even from its highest branch it cannot see the Persian Gulf, Bahrain's largest body of water.

***

I have met many people, even some of my family from ex-colonies themselves, who still like to see the British Empire as a version of Ninhursag, helping things grow in the harshest of environments. As we saw from the parliamentary transcripts, this is still held to be true by the political representatives of the British people with regard to the supposedly new role of 'Global Britain', and there have been few or no official challenges to this viewpoint. Not a single MP thought to question the use of imperial language during the discussion or the complexities of the history of Britain's presence in the region, beyond a repeatedly stated 'close relationship'. I think it must be possible to be self-critical and reflective, whilst also attempting to maintain a satisfactory level of economic security for the country in lieu of completely dismantling and rebuilding the system. But perhaps not. Capitalism is a system of colonisation, white supremacy and inequality, so whilst it is here, maybe it is impossible to expect a country to be financially stable and self-reflective, culpable and restorative. Is it possible to even expect this from one person in such a system? Can I hold myself to it? Can you?

IRAQ
IRAN
Bubiyan Island
Kuwait Bay
KUWAIT
Persian Gulf
SAUDI ARABIA

Tributaries

(Kuwait)

You must have been told what is expected of you, Sabrina?
We must be accountable, knowledgeable,
we must decide if your lies will be for us, or others.
It is this, more than anything else, which is of the utmost importance.

~~~

I have never been able to satisfactorily hold together the threads of what is expected of me, whether by family, friends, colleagues, clients, myself. Maybe the mess of mixes – class, culture, ethnicity, religion – makes it impossible to decipher the unspoken expectations emanating from each different hand I have to shake and cheek I have to kiss. Maybe I am just not skilled at reading people. Maybe on occasion they want too much from me, considering what it is they can offer, and so I feign ignorance, which is easier than saying: I deserve more than you can give.

Though these feelings have flowed since I was a child, it was during one of the MOD Fast Stream training programmes when they became impossible to ignore. These
~~~

programmes were broad-based and some were relevant to all departments, such as leadership and managerial courses and classes to learn about the workings of the House of Lords or the sprawling rules of Parliament. Others were specific to a department, and for the MOD this meant having a detailed knowledge of what the British military consisted of and what it did, including joining infantry training, flying on Chinooks and visiting arms factories.

As I've mentioned, I was the only person with Arab heritage in that year's cohort. I didn't meet another Arab person as we went across countries on our rickety minibus, from training ground to airbase and back again. Growing up in London and then working in its nightlife industry, I had genuinely never before been around entire groups who could be described for tick-box purposes as homogeneously white British. Even the Whitehall offices of the MOD hadn't shown me this. The furthest I had to go to find another person with a Muslim name was the desk in front of me. But on these training grounds, it felt that such a find would happen only if I had Top Secret access to the files on all the current targets. It was bewildering, trying to understand what I was doing there, why I had put myself there, even if somewhat indirectly.

I met children. Sixteen- and seventeen-year-old boys with machine guns, shooting on practice ranges in the beautiful Brecon Beacons. Ranges that were made to look like bombed-out villages. We hid behind concrete to smoke cigarettes whilst I asked them why they'd joined.

They looked at me, awkward in my helmet, fatigues, backpack, the gun I was given on entry banging my elbow every time I took a drag, and laughed – we should be the ones asking you that, sweetheart. I laughed too. They were right. My heart was full for them and for so many others I met. There was a conflicted admiration I felt for those who took such pride in their work, their daily purpose, their training. The ironically humble exceptionalism with which they regarded themselves, their colleagues, their country. Their absolute commitment to all of those things, or to the version of those things that they believed in so deeply. The camaraderie of belonging to a group who would kill and die for you, without drama or expectation of anything but the same from you. But their die-for list was full of people with names like theirs. Their kill list was full of people with names like mine.

~~~

During one presentation, we went into a simulation room. In my memory it is like one of those re-created
~~~

experience rooms in a museum – to feel what it was like to live in medieval London or pray in a Roman temple, artefacts moulded from polystyrene and painted in rough estimations of reality. Though the weapons lining the walls on this occasion were likely real. A TV was set up, chairs in front of it, cinema-style. Our course leader, perhaps an army general, more likely a captain, played a video to show us the prowess of the most recent tank purchases the British Army had made. The damage they could do, the destruction they could handle, the bravery of those within, their celebratory shouts as they emerged unscathed from showers of gun-fire in a scorching desert. The soundtrack to this surreal promotional film was Frankie Goes to Hollywood's 'Two Tribes'. The General/Captain happily sang along about the points that could be scored from two tribes going to war.

Being asked to admire an armoury made to kill people, at that time specifically mostly Muslim people, in Arab countries, pushed some kind of noise from my mouth. I don't remember what. It was enough for him to switch it off and look at me, ask if I disapproved.

– Yeh, I do, yeh.

It was all I could manage.

For the second time that week, I was asked,

– What are you doing here, then?

This time I didn't laugh.

~~~

I wanted to learn how to shoot. On reflection it was likely a belated desire for self-protection following my tumultuous teenage years. Maybe it was just the Americanisation of my imagination, along with my propensity to be attached in one way or another to the underworld of things. Whatever it was, the MOD training had given me guns and now I wanted to know how to use them. I just didn't want to learn how with people I thought might like to shoot me. During a few days of annual leave, I went to Kuwait for some shooting training with a private British company, run by ex-army members. It felt irrationally safer to be in a shooting range with a majority of Arabs, rather than posh white officers and poor white infantry kids. I was a decent shot. I enjoyed the disassembling and reassembling against the clock. That endless search for power, being momentarily found as cold metal gripped by my fingers.

When I left the MOD, I looked back on my time in Kuwait learning to shoot with ex-British Army members and the Kuwaiti Army, drinking Starbucks coffees with them whilst discussing the best pistols, going to
~~~

TGI Friday's for milkshakes to celebrate gaining my Beginner's Certificate, and I realised I was having some kind of breakdown. Able to emotionally disassociate so easily from what I was actively pursuing, which was to be a trained shooter with a British passport in an Arab country. Whilst my intentions were as wholly personal as anything can be, which is of course not a huge amount, the conscious-or-not political statement of my presence on such a course becomes glaringly apparent with the tiniest bit of historical context.

~~~

Simplest to start with Sykes–Picot: a secret British–French agreement made after the First World War concerning the dismemberment of the Ottoman Empire. This agreement will raise its Anglo-French head throughout the rest of this book, as it arbitrarily demarcated the borders of the region's countries as we have come to know them and remains absolutely pivotal to the region's disputes today.

In 1916, the UK and France agreed on which Arab countries each of them would control when the Ottoman Empire finally fell following the First World War, with assent given by Russia and Italy in exchange for the smaller areas they would also be apportioned. The Sykes–
~~~

Picot line literally divided much of the Middle East into A and B. France would 'control and influence' the A area, made up of modern-day south-eastern Turkey, northern Iraq, Syria and Lebanon. The UK would 'control and influence' the B area, made up of what is now southern Israel, Palestine, Jordan, southern Iraq and Kuwait. Russia would get Constantinople, the Turkish Straits and western Armenia. Italy took southern Anatolia. The 'Palestine Region' would come under 'international administration'. This pick'n'mix giveaway was drawn up by two men. François Georges-Picot was a French lawyer and diplomat who, prior to this agreement, had played a major role in the execution of Arab intellectuals in 1915 that was ordered by the last Ottoman military leader. Mark Sykes was a British baronet and politician-diplomat who had founded Britain's intelligence bureau in Egypt.[38] Together, Sykes and Picot would decide what lines millions and millions of future Middle Easterners would live behind. The whole thing was a disaster, which has been well documented and much lamented, particularly by Arabs. The details of the 'secret' agreement were revealed by the Bolsheviks after the Russian Revolution, when they came across the treaty one year after it was signed. The Allies were embarrassed. Especially Britain, as the treaty showed it had made contradictory and impossible promises to the Arabs, to the Zionists and to the Sharif of Mecca, Hussein bin Ali, whom it had gained support from to quell the Arab rebellion against the Ottomans by promising him an independent Arab state. Embarrassed, but not enough to rescind any of their 'power and influence', which at the very least can be credited with the dispossession and oppression of the Kurdish, Armenian and Palestinian people and, many would argue, every consequent and current conflict in the region.

Kuwait's borders were drawn so as to dominate access to the north-west of the Persian Gulf. Britain had a long-standing arrangement with Kuwait that allowed it freedom of movement in Gulf waters under Kuwait's jurisdiction. Therefore the British ensured Kuwait controlled as much of the coastline as possible, in order to maintain unhindered passage between the Persian Gulf and their power centre, India. This meant Iraq, despite being 2,360 per cent larger than Kuwait, was given next to no access to the sea. As we will come to see, Iraq's lack of access to the waters of the Gulf brought disastrous consequences which are still playing out today.

Contact between Kuwait and Britain had begun when Karim Khan of Persia captured Basra in 1776, meaning the East India Company had to make Kuwait, instead of Basra, the south-eastern end of its desert mail route to

Aleppo. But it was in the last decade of the nineteenth century that things began to heat up, with both the threat of a German railway and a Russian invasion jeopardising Britain's easy access to the region's waterways. Around 1886, Britain had become alarmed by Russian and German interests gaining ground in the Gulf, an area it was critical to control for intercontinental trade. Bahrain was the first Gulf country Britain had established an official claim on, but when the Germans began the Baghdad railway scheme in the 1890s with Kuwait as the terminus, Kuwait became of pressing interest to Britain, as well as Russia and Turkey, as it was a de facto part of the Ottoman Empire. Britain was determined to prevent Germany from controlling trade from the eastern shores of the Mediterranean to the Persian Gulf. Kuwait's ruler, Sheikh Mubarak, had already appealed to the British for their protection against being completely absorbed into the Ottoman Empire. Britain agreed, in exchange for the promise that the Sheikh would not sell or lease any land to another country without Britain's consent,[39] which would stop Germany's railway scheme going ahead as planned. Britain presided over many of Kuwait's foreign and domestic affairs from that point, including negotiating, diplomatically and militarily, access to large areas of coastal boundaries that had previously been regarded as

Iraqi and Saudi territory. Britain justified this in economic and moral terms, claiming that having more control over the Persian Gulf would not only enable it to fight off foreign threats in the area and maintain the smooth running of its major imperial trade routes, but also to combat slavery and piracy more effectively.[40] There were multiple altercations both diplomatically and militarily with the Ottoman Empire over Britain in Kuwait, until 1913, when the Ottoman Empire was already folding and Said Halim Pasha, the next vizier, wrote in his diary:

> We would be in conflict with England over two desert districts like Kuwait and Qatar. What benefits could we hope for from these unimportant lands. I decided it was better to leave Kuwait and Qatar to England and concentrate on the rich provinces of Iraq. On Friday 23 May I came to the Ministry of War. I looked through the dispatches sent by Hakki Pasha [the Ottoman negotiator] from London. We would extend the Baghdad line only to Basra; [under British pressure] we had abandoned the plan to extend it as far as Kuwait. Hakki Pasha noted that an agreement had been reached with the English about such small bits of our empire as the sheikhdoms of Kuwait, Qatar, and Bahrain. We had also resolved the question

> of traffic in the Persian Gulf where England wanted to increase her influence. Even though England's [growing] influence in the Persian Gulf is against our interests, it is necessary for us to remain silent in order to get along with this state . . . No Turkish troops will be stationed in Kuwait.[41]

These 'unimportant' lands soon became black gold mines for the British and for those Kuwaitis and Qataris who were in Britain's favour when oil was discovered, first in Kuwait, in 1938, and then in Qatar, in 1940, marking an important shift in how Britain regarded these countries and their rulers.

~~~

Oil money also meant an important shift in Kuwaiti women's access to education and their involvement in public life. Or, perhaps, a slow return to how things had been before the British and Europeans had come to Arabia. Prior to the eighteenth century, women's contributions to Central Arabian society – culturally, religiously, medically and militarily – were recorded by Muslim historians, with far less prominence than men's, but ever present, from AD 1233 to 1875. After that, these sources become far more difficult to find.[42] Perhaps it
~~~

is a coincidence that this erasure of women and their contributions coincided with the arrival of the British and their economic model based greatly on societal control and patriarchal hierarchy. Perhaps not.

Either way, in the late 1800s of pre-oil but colonially-attached Kuwait, women could no longer move about as freely as pre-colonial Arabian texts suggest some did previously. Poorer women had to leave the house to work and fulfil the needs of wealthy families, whilst wealthy women were confined to their large homes. In 1937, a school for girls was finally established, but it had no students for almost six months.[43]

These attitudes changed after the accession of Emir Abdullah al-Salim (1950–65), whose significant benefits system started flowing oil money to full Kuwaiti citizens; a system that continues to this day, though with growing trepidation as oil prices continue to drop.[44] As families began to accumulate wealth, they would send sons, and occasionally daughters, to be educated abroad. They returned full of ideas and ambitions inspired by reading the Egyptian writers and activists of the time, including educating women. These graduates began to set up various print media projects, making space for their sisters and nieces to be involved in them, too. Whilst the opportunities to write for the 'Women's Corner' in magazines mostly applied to wealthy women, it was the first time female education, veiling and paid work had been discussed publicly by women and it shifted the discussion on their participation in a modern Kuwait.

In 1962, Kuwait's constitution was drawn up by Egyptian jurists – it had been Egyptian jurists who created the first Arab civil court in the 1870s – and comprised a mix of Islamic, Egyptian, French and British codes and laws. Whilst equal rights were promised in the constitution, it made the family rather than the individual the basic unit of Kuwaiti society, and women were still culturally expected to raise a family and be domestically dedicated, as was the case in most places at the time. Despite increases in family allowances, compulsory education provisions for girls and boys, and universal healthcare, women were still excluded from political life and not permitted to vote. This didn't change until 16 May 2005, after forty years of suffragist struggle.[45] Lulwa Al-Qatami was a key leader in this suffrage movement as Kuwait's most prominent feminist and head of the Women's Cultural and Social Society of Kuwait, which she co-founded in 1963. The Society made huge improvements in the population's literacy and helped improve the education and care of children across the Arab world, as well as staging numerous non-violent direct actions, such as turning up to vote at every election for decades, even though they were refused. Unfortunately, all the records of their society were destroyed during Iraq's 1990 invasion of Kuwait. This is a recurring problem in the region: archives are difficult to find due to invasions, occupations, thefts and colonisation. This loss perhaps also partly explains the lack of contemporary documentation. To find out about key Kuwaiti women from the 1960s I was grateful to finally come across a video featuring interviews of the women I mention below, as well as Lulwa Al-Qatami, because there were very few other sources that celebrate and document these trailblazing women.

Fatma Hussein was the first woman to produce a television show in Kuwait, one which became pivotal for similar programmes examining family life across the Arab world to be seen as 'family shows' rather than 'women's shows'. *Dunya Al Usra* (*Family Life*) ran for fifteen years, and Hussein says her objective with the show was always to fight for women's rights in every aspect of their lives. Ghaneema Al-Marzoug was the first woman to be a licensed publisher and to be editor-in-chief of a magazine, as well as being the creator of a publication concentrating on women's issues and family, not only in Kuwait, but across the Gulf States. Mariam Al-Ragom was the Kuwaiti nurse who set up the region's

Society for Nursing and changed the profession for the generations who came after her.[46]

~~~

And there I was, dressed in fatigues and holding sniper rifles at a shooting range in Kuwait City, just a few years after these women had seen their lifelong commitment and dedication realised in political terms, aiming bullets at paper targets and feeling sorry for myself because I wasn't being accepted as all the things I was, by an establishment that had never said that it would.

An establishment that still makes a serious percentage of its GDP from this centuries-old water pact with Kuwait. Though unhindered access to the Persian Gulf is still a top military and economic strategy for the British, it is Kuwait's immense market for arms that is now the government's main financial focus. When I was being timed disassembling a Glock handgun in a Kuwaiti shooting range, I didn't know that Kuwait was one of the UK's biggest arms customers, averaging an $800 million spend annually.[47] Saudi Arabia, Qatar, Oman and the UAE are other Middle Eastern countries that have helped the UK to recently become the second-biggest arms dealer in the world after the USA. In 2020, the UK had a 16 per cent share of global arms exports, to America's 47 per
~~~

cent, Russia's 11 per cent and France's 10 per cent. Their contracts worth £100 billion over the last decade were in large part thanks to Typhoon fighter jet orders from Kuwait, Qatar, Oman and Saudi Arabia.[48] As mentioned, the UK resumed arms sales to Saudi Arabia in 2020, even after the Court of Appeal had ruled the sales illegal due to evident human rights violations by Saudi Arabia against Yemen, where the arms sold by Britain are currently being deployed against civilians.[49]

An establishment that gave the huge Persian Gulf the nickname 'the British Lake' due to their historic domination over it,[50] and now maintains its position as the sixth-largest economy in the world by selling the countries surrounding that 'lake' weapons made by the British, to kill each other with.

An establishment that drew the modern lines of Kuwait and Iraq to best serve itself, so that in 1990 Iraq's Saddam Hussein invaded Kuwait, beginning the first Gulf War. Kuwait holds around 7 per cent of global oil reserves, making it the fourth-largest reserve in the world, producing 4 million barrels per day as of 2020, whereas Iraq is behind it as the fifth, despite its much larger size.[51] Oil was one point of many areas of contention, as Kuwait began to increase production, which affected Iraq's oil revenue, and Iraq accused Kuwait of stealing its oil from across the border. Other motives are speculated to be Iraq's inability, or unwillingness, to pay Kuwait the $14 billion it had borrowed to fund its war with Iran, as well as the simmering resentment Iraq had at Kuwait's British-drawn access to the coast, causing it to be an almost landlocked country. Iraq claimed that Kuwait had always been an integral part of Iraq, prior to British interference, even when under Ottoman rule. Within two days in August 1990 Iraq had annexed Kuwait, leading to a seven-month occupation that only ended after the biggest coalition of superpowers since the Second World War joined forces. Led by the US, a coalition of thirty-five nations invaded Kuwait, and the following conflict became known as the first Gulf War, marking the first time live news broadcasts were made from the battlefield. After just over a month of fighting, resulting in estimates ranging from 8,000 to 50,000 Iraqi military deaths[52] and around 100,000 regional civilian deaths,[53] versus 392 total deaths in the coalition forces, Iraq withdrew its claims on Kuwait.

The coalition forces stopped the Iraqi occupation of Kuwait, but had not solved the underlying problems, arguably laying the groundwork for even more severe ones, as the defeat plus the near-total trade and economic sanctions that were placed on Iraq by the United Nations took horrific tolls on the Iraqi economy and consequently every aspect of society. Amongst the consequences of this humanitarian crisis, there were an estimated half a million child deaths, caused directly by the sanctions.[54] As in Yemen, the poverty, difficulty of life and lack of access to resources contributed to the growth of existing sectarian and ethnic feuds. In 1988, Saddam Hussein had used nerve agents, mass executions and torture against the Kurds, killing an estimated 100,000 people.[55] This horrific genocide strengthened the urgency of the Kurdish cause for autonomy and, in the power vacuum left after Iraq's defeat in Kuwait, Kurds and Shi'a groups, who were also systematically targeted by the regime, led an uprising against Saddam Hussein, with Shiite rebels taking control of the city of Karbala. The regime once again responded with the use of chemical weapons.[56] These attacks made Hussein the first leader since Hitler to use chemical weapons against his own people and to deploy chemical weapons on the battlefield. The international community was hugely concerned, and in 1992 the UN finally banned chemical weapons under the Chemical Weapons Convention. Hussein regularly refused access to UN weapons inspectors and by the late 1990s had become the world's most renowned dictator, for his brutal genocidal actions and defiance of international law. In 2003, following 9/11, a disastrous invasion of Afghanistan and an abstract 'War on Terror' to win, America knew it had power to prove, and a beleaguered, feared Iraq was seen as the place to prove it.

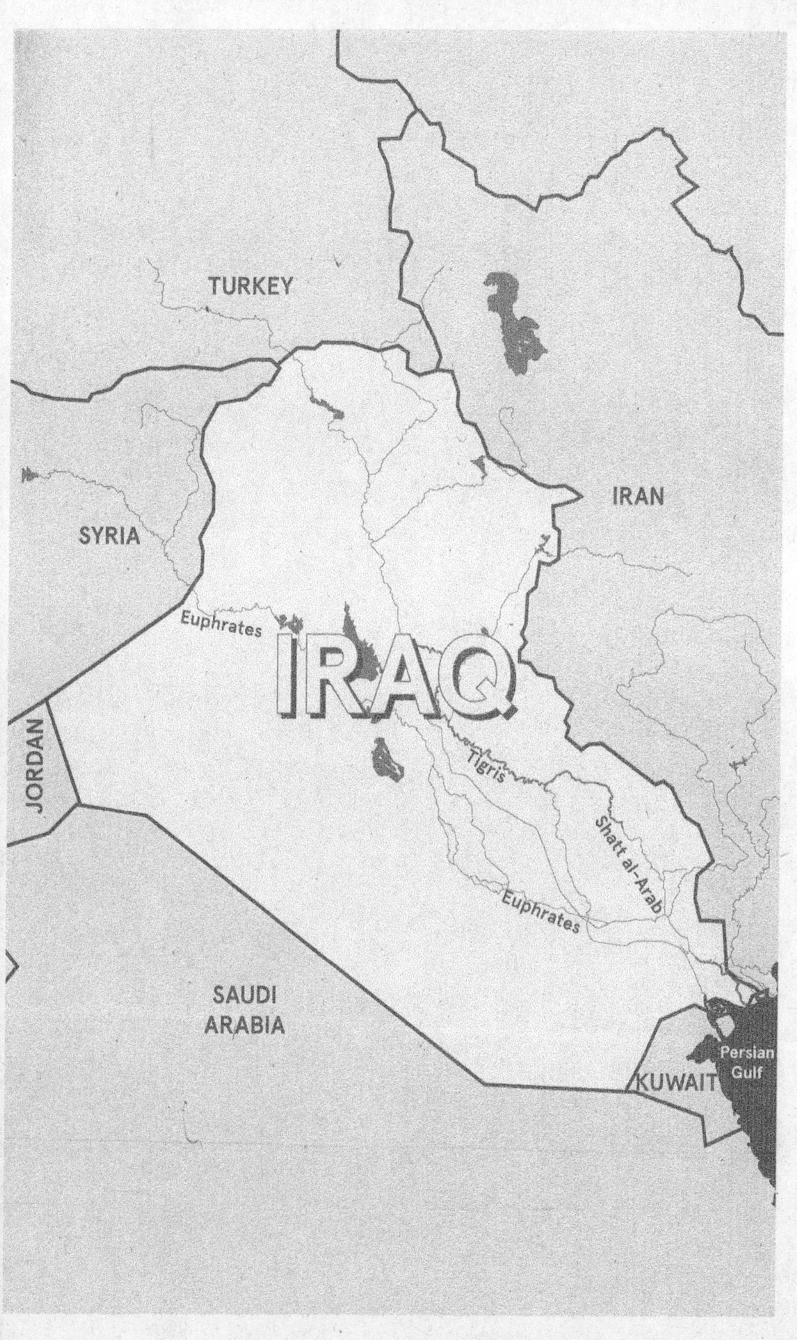
TURKEY
IRAN
SYRIA
Euphrates
IRAQ
JORDAN
Tigris
Shatt al-Arab
Euphrates
SAUDI
ARABIA
Persian
Gulf
KUWAIT

Drainage

(Iraq)

Do you consider yourself an honest person?

~~~

– Hmm. Well, I have to address your debts. Credit cards, overdrafts and one that you omitted from the Developed Vetting form entirely – your student loan. I certainly wouldn't call that particularly honest, would you? And why would you have such a large student loan when you were a recipient of a university Hardship Fund loan as it was?

– That's the thing about hardship, sir, it's hard.

– We can't satisfy this process with adverbs, Ms Mahfouz, I need concrete examples. Specifics. Specificity is everything.

– And if I give you some, what does that do?

– It mitigates the risk.

– What risk?

– The risk you pose.

– Due to my 'hardships'?

– Look, we must have due diligence, we are the
~~~

intelligence community. Bribery is the biggest threat to our national security –

– I thought it was Al-Qaeda?

– *Yes. But how do they threaten us? How do they infiltrate, find out our movements, block us, throw obstacles that stick? Any individual who is granted Top Secret Clearance, who has access to the kind of information that is handled at the highest levels of government, must be vetted to an extent that we can confidently say there is next to no risk that this individual would be susceptible to bribery and therefore treason. It's perfectly reasonable. Debt, skeletons in the cupboard, certain family histories, political leanings – you must see these need to be the first things we address. And I am struggling to see how you managed to accrue debt, despite working in stripclubs and nightclubs and receiving Hardship Loans. Can you see where I'm coming from?*

– I do. I do. Okay. As a teenager, I stripped for money. In stripclubs. I mean, like literally took off my bra and knickers in exchange for twenty-pound notes which paid for my home and food. Not because I thought it'd be a rebellious, sexy thing to do but because if you want to get an education outside of what has been mapped out for you, it's

a shock of a cost and a culture and London's a
vulture, as I'm sure you know.
But it was good, in its way. I came from a
relationship, a long one, a bad one. He used to do
all sorts of – unspeakables, really, even though I'm
trying to speak as openly as I can with you, sir.
I tried to buy him, his gentle side, he had one.
Tried to pry it out with gifts. It was the start of
the credit card boom. I applied for them all, they
gave me thousands, despite being eighteen, despite
signing on, despite retail and McDonald's making up
the chunk of my CV. I bought him YSL shirts and
Ralph Lauren polos. Reebok classics in every colour.
Laurent Perrier champagne and grams of coke. Each
gift gave me the gift of another day alive. I don't
regard that as an exaggeration.
Until I managed to leave. Finally, painfully by the
very literal skin on my teeth. After escaping that.
Well. For all its bullshit, stripping *was* empowering,
it was absolutely amazing, spiritually, emotionally,
physically healing. Healing more than any therapy
would have been even if I could have afforded it.
More than the drugs they tried to give me at the
GP when they said the waiting list for talking was
longer than they trusted me to stay quiet.

Healing. Taking money from a man who just watched, couldn't touch, after years of fearing even the little fingertip of someone who was supposed to love you.

I cannot be bribed by anyone for what I did. I cannot be shamed for it. Stripping . . . it might not have cleared my debts or given me enough to get to uni without a huge loan and hardship funds, but well, it saved my life.

Is that specific enough?

– It'll do, for now. Thank you, Ms Mahfouz. But I do reiterate, this process must be based on trust. And you must see that sex work in one area will raise questions about whether you engaged in other areas, and if you did, there is no judgement, I repeat, no judgement. We just need to know, so we can accurately jud – access – your risk levels. So please, trust me, tell me.

~~~

Money and honesty. A tricky combination. If a global monetary system is based on dishonesty, then how on earth can an individual make any of that money in a way that could truly be considered honest? A discussion on class and capitalism was not something the interrogator was willing to entertain. As the years have gone by, I'm
~~~

particularly surprised by that, as any radical thoughts on class and against capitalism are often treated as the most dangerous ones, to be suppressed immediately. I suppose he didn't need to engage in a discussion with me to see where my views lived with all that. Every decision I have ever made is inextricable from my class in this country and perhaps from the fact that this class changes when in my other country, Egypt. Frantz Fanon explored this in *The Wretched of the Earth*, examining how the social classes of the colonised world do not correspond to those of the colonial world and the alienation this can cause. Unionised trades in Europe like train drivers and nurses are considered the middle classes in countries such as Egypt. My dad, whose father owned his own successful business, grew up as middle, then upper middle class in Cairo. He inherited the company and so maintained that class status, which as his child I then occupied when I lived there. However, exchanging the money made from that business into the currency of the old colonial power when living in the UK put our family into poverty. Without any other British class-defining factors such as higher education, familial legacies and professional skill sets at our disposal, and with the only British strand of the family being poor and working class, this was my status, regardless of how I was seen and the place I occupied in my other home, an old colony. Although Fanon meant this for those in the colonised lands, I have felt it throughout the diaspora of colonised people now living in the coloniser's headquarters:

> . . . that horde of starving men, uprooted from their tribe and from their clan, constitutes one of the most spontaneous and the most radically revolutionary forces of a colonised people.[57]

We are now starting to see this force's growing strength in the UK, with campaigns such as the NUS's Decolonising Education and the Museum Association's Decolonising Museums campaign. These campaigns and others have found widespread support, partly thanks to social media, where there has also been an increased sharing of academic books and articles that look critically at the empire's legacy and writing that examines the more personal experiences of being from the colonial diaspora, both fictional and factual. These resources have prompted a nascent mainstream understanding of colonial legacy, white supremacy, and the systemic racism, sexism and ableism that frames a society which has been built and sustained to benefit those at the invented 'top' of it. From these campaigns to decolonise the curriculum to the toppling of statues, the work that post-colonial academics – primarily in the global south – have been doing over the last few decades can now be felt in the arts and popular culture more palpably than I can ever recall. Netflix's most popular show of 2021, *Lupin*, made by French and British creatives, featured French West African actor Omar Sy as the protagonist, a 'gentleman thief' in Paris, originally from Senegal. His story arc is inextricably linked with European colonialism, even if it is never explicitly stated. The top-selling fiction in the UK in 2020 included the Booker Prize-winning *Girl, Woman, Other* by Bernardine Evaristo,[58] which delves into the histories of colonialism and migration that brought the twelve women protagonists or their ancestors to Britain. One of the non-fiction bestsellers for 2020 was Reni Eddo-Lodge's seminal *Why I'm No Longer Talking to White People about Race*,[59] which centres on systemic racism and white privilege in Britain, both of which were essential for empire-building and control.

Despite the distinct lack of Middle Eastern and Arab voices platformed amongst this mainstreaming of cultural decolonising, it's a huge and hopeful push forward. It is not a given that this will continue though; it requires constant effort and vigilance to ensure we do not allow a narrow, nationalistic narrative to re-emerge as the default. We are at a time when British people are so used to being lied to by the leaders of the country, that the irrefutable exposure of deceit no longer merits a call for resignation, even from an opposition leader. This is a culmination of decades of government lies to the public with zero consequences for the perpetrators, no matter what devastation the dishonesty caused.

~~~

One of the biggest of these lies ever told was regarding the Middle East: 'Weapons of Mass Destruction have been, and more will continue to be, found in Iraq.'

So said the spies, intelligence on both sides of the water, in the US and UK. Going to this landlocked country they set up in the first place, to set it up once again with a string of spectacular omissions, never quite admitting the so-called 'sexing up' of documents that justified the unjustifiable. But how did it even get to that point and why did the US and UK need to bend so many truths when they already had the set-up to get what they wanted from this oil-rich nation.

After the end of the Cold War, America needed to demonstrate its now-unilateral global power, show the world it was as Great as its own old coloniser, Britain, used to be, and the perfect place to do that, following the
~~~

war with Kuwait, and the global villain status of Saddam Hussein, would be Iraq.

In 2003, Bush and Blair invaded Iraq. Bush with 130,000 soldiers, Blair with 45,000 soldiers. They did this despite receiving no backing from the United Nations Security Council, a rebellion from eighty-three Labour MPs (though 412 votes for to 149 votes against did give Blair overwhelming support from the government for military action in Iraq, based on the WMD claims) and a 2-million-person march against the war in London, with millions of others joining in around the world. The invasion was being justified by claims from the US and UK intelligence agencies that Iraq had stockpiled weapons of mass destruction and could deploy them within forty-five minutes. Claims that a British Defence Intelligence weapons expert, David Kelly, revealed were highly doubtful in May 2003. David Kelly was found dead in July 2003. In January 2004, the Hutton Report into his death cleared ministers of wrongdoing and concluded Kelly had taken his own life. Six months later, the Butler Review on military intelligence backed up David Kelly's concerns, revealing that sources were not checked thoroughly and claims of Iraq's forty-five-minute WMD deployment abilities were unsubstantiated. Despite these revelations, the occupation of Iraq by UK forces continued until 2009 and by US troops until 2011. The removal of Saddam in 2003 and his execution in 2006, along with the US ban on the long-ruling Baath Party and the forced disbanding of the military, left the country open to sectarian and political violence, resulting in a civil war and, by 2014, ISIS controlling a third of the country. Although Iraqi forces regained control over all territories from 2018, there are still constant threats from militias and non-state actors, partly as a result of the UK/US destabilising the area over the last few decades. In January 2021, there were still 2,500 US troops in Iraq[60] as well as a newly enlarged NATO force of 4,500 and 400 UK troops, all of whom are officially there for training the Iraqi military and security forces. This number doesn't include personnel from Private Military and Security Companies (PMSC), because although modern warfare would arguably be impossible without their involvement, these companies are not subject to checks and operate largely outside of any scrutiny. The UK is the global centre for PMSCs and many are run by ex-Special Forces from the British Army.[61] Airstrikes by the UK and US are still commonplace in the country, with the justification that they are helping the Iraqi military fight against Islamic State forces.[62]

~~~
~~~

Iraq is where the first recorded water war in history took place. Around 4,500 years ago, the Sumerian city states of Lagash and Umma, located in modern-day Iraq, went to war over the waters of the River Tigris. The King of Lagash had destroyed a canal to divert it into his region, which deprived Umma of fresh water, essential to agriculture and survival, especially with little rainfall in the region. The two cities eventually made what is considered to be the first boundary water agreement, the Mesilim Treaty, and since then there have been no wars anywhere that can be explicitly called 'water wars', even if water tensions and disputes have contributed to wider decisions to wage war. In the early 1990s, after the shattering defeat of the first Gulf War, Saddam Hussein famously drained Iraq's southern marshes, thought to once contain the biblical Garden of Eden and now a UNESCO World Heritage site. The once-biodiverse wetlands are fed by the Tigris and Euphrates rivers. Hussein drained the wetlands in order to force out the Ma'dan people, or Marsh Arabs, with water diversion tactics, for their supposed treachery in the Iraq–Iran War of 1980–88. Hussein was clamping down even further on any potential dissent following the coalition victory in Kuwait, and the 1991 Shi'a and Kurdish uprisings that had taken place during a ceasefire in the war. The official reason given for the draining was to rid Iraq of a breeding ground for mosquitoes and to reclaim land for agriculture. Over 200,000 Ma'dan people were displaced and the consequent level of state-sponsored violence against them has been called genocide,[63] whilst the draining of the marshlands has been described by the United Nations as an environmental catastrophe on par with the deforestation of the Amazon rainforest.[64]

The Iraq–Iran War itself can be linked to a river dispute, the long-running Shatt al-Arab border issue between the two countries, which in turn can be linked to British interests in the area. All things which are explored later in this chapter. All things I was unaware of when the interrogator quizzed me once again about my generous Iranian friend who had helped me rent an apartment in Mayfair, once again requesting honesty but not accepting it as such, framing questions as if they were concerned with morality, or even fairness, when really the incredulity was all down to gender economics.

– You can't honestly expect us to believe that your Iranian 'friend' let you rent an apartment in his Mayfair building for the same cost as if it were in Lewisham without any sexual relations existing

between the two of you . . . or in exchange for any other kind of . . . service?

Why would I not expect them to believe that? I couldn't see how he saw my place within that world then, I was so close to it. I've gained some pretty good ideas of how I must've been regarded as the years have gone past and my distance from the London nightlife world has grown.

Meeting customers whom you serve bottles of champagne to whilst wearing a corset and stockings, laughing about their taste in shoes as they ask you the name of the stripper they've been eyeing up and then becoming friends with them was not unusual. But to become great friends with one of them, so much so that they launch you into more aspirational versions of yourself and literally give you the keys to leave a tower block, but never let that define the friendship, was unusual for any workplace, let alone a stripclub. To try to get the interrogator to understand a bond of friendship he could only see as suspicious, I first had to try to explain the job.

After realising the waitresses in the stripclub where I worked made the same money as me or more than I did but didn't take their clothes off, got more free drinks and even read their books sometimes if their tables left early,

I started working as a waitress rather than a stripper in Stringfellows. The money was better, I got shift pay and didn't have to pay to be there, as all strippers do. I wasn't under pressure to do at least five dances to break even. I stood along the wall with my tray at the ready, waiting to be beckoned by the men at the tables, who weren't encouraged to go to the bar – that was only for the broke customers. I got to chat with the bouncers and the other waitresses, the barmen, the managers. Once I was settled in, I would write notes on my order pad until I was summoned to a customer, the beginnings of my writing life right there under the hazy dry-ice disco lights. But I also got inspected, head to toe. No runs in stockings, my heels high enough, my knickers lacy enough, my corset fitted enough, my nails filed and painted, my make-up tasteful, my hair either neat and practical or wild and sexy. Every night we were lined up and either given the nod or instructions on what to rectify before we could begin our shift.

Into all this walked many men who became, what I would consider at the time, good friends. Who got to know about the essays I was writing for uni, who tipped me enough to cover my rent, food and travel in one night, who gave me contacts for work experience and career advice. Perhaps befriending me lent some legitimacy to their visits. They weren't just coming to see naked women, they were coming to catch up with mates. And on rare instances, those friendships became real enough to exist outside of that surreal subterranean world where we had found some kind of comfort in each other. My Iranian friend was one of those rarities, one who gave me a home on one of the country's most famous streets for the same price as my Lewisham council flat. He wasn't a huge deal older than me. He'd been left money, I think. Asking about sources of wealth was something I'd been trained out of by stripclub etiquette, and it no longer felt strange to not know how someone so young had a Bentley and a Rolls and no idea of how much a loaf of bread might be. We would party on my nights off, or after work, in nightclubs and hotels. Friends from all parts of my life would come, he'd get them anything they wanted. My family came sometimes too, my dad, my aunty. He became family, in a way. After a few months of hanging out regularly, he took me and a group of our friends to his penthouse apartment overlooking the MI6 headquarters and as we sat next to the floor-to-ceiling glass, the little grey waves of the Thames sparkling in the corner of my eye, he asked me if I had feelings for him. He explained that it often happened when he became good friends with a girl, she would mistake his generosity for something romantic and it would get awkward. He didn't want that to happen with me. Maybe it was a ruse to avoid rejection if he did have other plans. Maybe it was his way of saying he never wanted to get with girls, despite his proclivity for stripclubs. I was drunk and amused. I liked guys with more edge, I told him. Money is nice, but I need danger. We cracked up, he hugged me and that was that, never anything but friendship for years. I can see how no part of this might point to my suitability for a job in national security, but at the time I just thought it was living the life you had the best way you could.

~~~

For my interrogator, there had to be more to it than my friend being a wealthy man I met in a stripclub. If he had been an English earl (of which there had been many amongst the club's clientele), then I wonder if the problem would have required so many hours of examination. But he was Iranian. I'd never properly educated myself on the British history with Iran. After this questioning, I did. It was long, messy, bloody and full of ongoing suspicion and tension, all linked to water on some level, with Iran having double or more the amount of coastal access to the old 'British Lake' than any of the other six countries around it. But let's follow the river and look
~~~

at the Shatt al-Arab, which forms the boundary between Iran and Iraq before flowing into the Persian Gulf. It was this river that was at the centre of the first water-related tensions between Britain and Iran and, consequently, Iran and Iraq. The Shatt al-Arab, the River of Arabs, is formed by the confluence of the Euphrates and the Tigris rivers in Iraq. The Sykes–Picot Agreement formalised the river as the boundary of British influence and control in Iraq. Whilst this had always been important to the British due to the major trade route from the river to the Persian Gulf and to India, the discovery of oil and subsequent construction of the Anglo-Persian oil pipeline that ran along its banks made it even more vital to secure the region. The Anglo-Persian Oil Company (now known as BP) signed a deal with the British Government in 1914 to supply the British Navy with 40 million barrels of oil over the next twenty years in return for £2 million and a majority shareholding. Six weeks later, the First World War began and that British Navy oil supply, and the river that held its pipelines, had never been more important.[65]

Over one hundred years since the Sykes–Picot Agreement was enforced, the river still constitutes the British-made border between Iraq and Iran on the last fifty miles of the river and continues to flow down to the

Persian Gulf. Being the only access point for Iraq to the Persian Gulf, the Shatt al-Arab has a strategic importance for the country's transportation and exports. Given the dry and humid climate, the water from the river is also crucial for agriculture. Although Iran has other points of access to the Gulf, a high quantity of crude oil produced in Iran continues to be transported through the Shatt al-Arab, from 1914 to this day.

Iraq always refused to recognise Iran's sovereignty over one half of the river, so in the 1960s Iran started scaling up its military capacities in order to defend its rights. In 1969, when Iraqi leaders claimed sovereignty over the entire river, Iran provided financial and weaponry support to the Kurds who were seeking autonomy in the northern part of Iraq. Knowing the Kurds were now supported by Iran and therefore supporters of Iranian policy towards Iraq, Iraq was forced to recognise Iran's sovereignty claims over the river, as they did not want to risk an armed Kurdish uprising financially and logistically supported by Iran.[66] This resulted in the 1975 Algiers Agreement, but Iraq withdrew from the agreement in 1979 as soon as Iran scaled down its arms support to the Kurds. Tensions escalated once again and became a major cause for the break-out of the 1980–88 war between the two countries, in which over a million people are estimated to have died and in which Iraq used chemical weapons against Iranians and Iraqi Kurds. The long-standing sectarian feuds, intensified by the 1979 Islamic Revolution in Iran, are cited as the main causes of the war, but Hussein declared the Shatt al-Arab dispute as the official justification for invasion and it certainly continued to be a major factor in the continuing conflict, including the territorial dispute over three Persian Gulf islands seized by Iran in 1971 and Iran's diversion of the Shatt al-Arab's tributaries upstream. During the war and its aftermath, at least 1 million people were displaced, including my friend and his parents, who relocated to London.

In 1988, after eight years of conflict, the UN brokered a ceasefire which both parties accepted. Though the war had formally ended, Iraq and Iran both publicly stated that their stance had not changed regarding the status of the Shatt al-Arab. The two countries didn't restore diplomatic relations for another two years, until after Iraq invaded Kuwait in 1990. At this point, Iraq had lost the support of most European countries and of America, who had all tacitly supported Iraq during the Iran-Iraq war, although Britain had also supported Iran's claims over the river in order to ensure its BP oil fields were protected. Conversely, Iran had managed to be on decent diplomatic terms with France, Italy and Germany by 1989. Hussein feared an alliance against Iraq, and finally withdrew all Iraqi troops from Iranian territory and agreed to divide the sovereignty of the Shatt al-Arab. After Hussein was captured in 2003, relations with Iran tentatively improved, though it took until 2019 for the 1975 Algiers Agreement to be reinstated.[67] Tensions remain, however, due to Iran's upstream water development projects, which cause major problems for Iraq's agriculture downstream, where the river is southern Iraq's lifeline.[68] It is also severely polluted, partly from the 'warship cemetery' that the river became during the war and also from numerous contemporary sources. A study published in *Nature Research* in 2020 found elevated levels of heavy-metal contamination such as nickel, chromium, manganese, molybdenum, copper, lead and zinc originating from fertilisers and sewage sludge, metal-welding workshops and the oil industry.[69] Iran and Iraq have somewhat surprisingly pledged to work together to clean up the environmental damage done to the river. Whilst progress remains to be seen and locals protest daily at the continued pollution of their water supply, the unexpected cooperation of these two countries can perhaps be seen as one hopeful indication of how the urgency of water issues in our current climate emergency might create diplomatic and environmental collaboration rather than further conflicts.[70]

~~~

Conflict is constantly associated with the Middle East, perhaps more than any other concept. This is part of Orientalism, hundreds of years of Western stereotyping of violent, unstable populations, even if simultaneously exotic and fascinating. These stereotypes have made it easier for Western powers to justify invasions, sanctions and occupations. Our discussion and representation of lives and cultures is never free from complexity, even if they are our own. But when a culture is associated so deeply with conflict, that complexity deepens as it overwhelms anything else about the place or the culture and it becomes difficult to not also centre the ones it is engaged in conflict with. As Ala A. writes in their article *Reimagining Iraqi History* (Tribes Post, November 4th 2020): 'In his seminal work *Orientalism*, Edward Said makes the distinction between latent and manifest Orientalism, and shows how the former is much harder to detect, as it is sewn into the very fabric of European language and culture. I think it goes without saying that no approach to Iraqi history, no matter who is conducting it, can be carried out without deconstructing latent
~~~

Orientalism constantly.' There is latent Orientalism in my own reimagining of Iraqi lives, despite my Middle Eastern heritage and lived experience. A few years ago, I wanted to write a play based on an Iraqi Botox doctor, as my Kurdish Iraqi friend and colleague Triska Hamid had recently written an article about Iraq's trend of tattooing eyebrows[71] and during her research had learnt that Iraqi women were the biggest consumers of aesthetic Botox relative to population at the time.[72] Considering the country had been decimated by false proclamations on the proliferation of WMDs that included botulinum, the base toxin for Botox, the subject stood out to me for further creative interrogation. I began the play, *Battle-face*, in a way which I hoped was, amongst other things, deconstructing Orientalism . . .

***

1.

Camilla (a journalist) and Ablah (a cosmetic doctor specialising in facial rejuvenation) are having an inter-view chat in a spare room at Ablah's clinic.

ABLAH: I'd estimate you're thirty-three years old, from the depth of the fountain of lines between your eyebrows. You take your job extremely seriously, working until the light late hours – revealed by the shade of dark skin under your eyes. You haven't been joyously happy for a while – the laughter lines around your mouth don't match your age. You don't eat well. You drink too much coffee. It gives you palpitations, but you drink it anyway because of this dedication to your work, again. And there's something else, something I can't quite put my finger on.

You'd have to sit under my lamp for a proper analysis.

CAMILLA: Wow. That was – amazing. I feel . . . naked.

ABLAH: Accurate, then?

CAMILLA: I had no idea all that was right here, on my face.

ABLAH: Most don't.

CAMILLA: So you really are the best.

ABLAH: Well, no – maybe, one of.

CAMILLA: Why do you do what you do, Ablah?

ABLAH: I love it.

CAMILLA: What exactly do you love about it?

ABLAH: The possibility.

CAMILLA: Possibility?

ABLAH: When a client comes to see me, they're hoping to rediscover their possibility. It's a beautiful thing to be able to help them do that.

CAMILLA: How *do* you do that?

ABLAH: I allow time to be pulled back inside a person's being.

CAMILLA: Quite a feat.

ABLAH: When they look in the mirror, they see no trauma, no disappointment, just –

CAMILLA: Possibility?

ABLAH: Exactly.

***

What I didn't realise at the time was that the framing of the play had already failed the deconstructing Orientalism challenge I had set myself. Although we see Ablah, an Iraqi woman at the top of her profession, excelling in her knowledge and power, what happens after this is that the 'journalist', Camilla, is actually a British intelligence officer, actioning a suborned recruitment attempt. The rest of the play is a dance between the two, the final steps left ambiguous. Contrary to my thinking at the time that this could be a subversive use of Orientalist tropes, it was a representation of an Iraqi existing in relation to violence, specifically British violence, throughout the play, despite the centrality of the character's profession and personal life.

~~~
~~~

We must keep testing, pushing, reassessing. Lend our support to groups such as the Iraqi Diaspora Creatives Network, created by Australian Iraqi artist Hajer, in direct response to a British TV show set in Iraq, *Baghdad Central*. Although it was widely acclaimed by non-Arab critics for centering an Iraqi perspective, something which has remained elusive across English-language storytelling, it did not have any creatives of Iraqi heritage apart from an associate producer, who Hajer says was brought on late in the process. This may not have been as frustrating if Iraq had a different relationship with Britain, but as it stands, Hajer knew she had to generate a space for Iraqi creatives that they would not otherwise be given. Her own words say it best:

> Iraq is one of those places that has had so much foreign intervention and been puppeteered for so long that a big part of what people know about Iraq and Iraqi culture is what's happened TO us – not what we've chosen for ourselves. Everything else has been stripped away from our stories as a result. Reclaiming the agency of our narratives is to me the only way we can heal.[73]

Healing requires honesty and, in a way, honesty compels the hard work of healing to begin both on a personal and a socio-political level, even if it wasn't part of the original plan. Perhaps that is why my interrogator refused to accept my honest answer to his questions. It is far easier to work with lies, especially when you have so much practice.

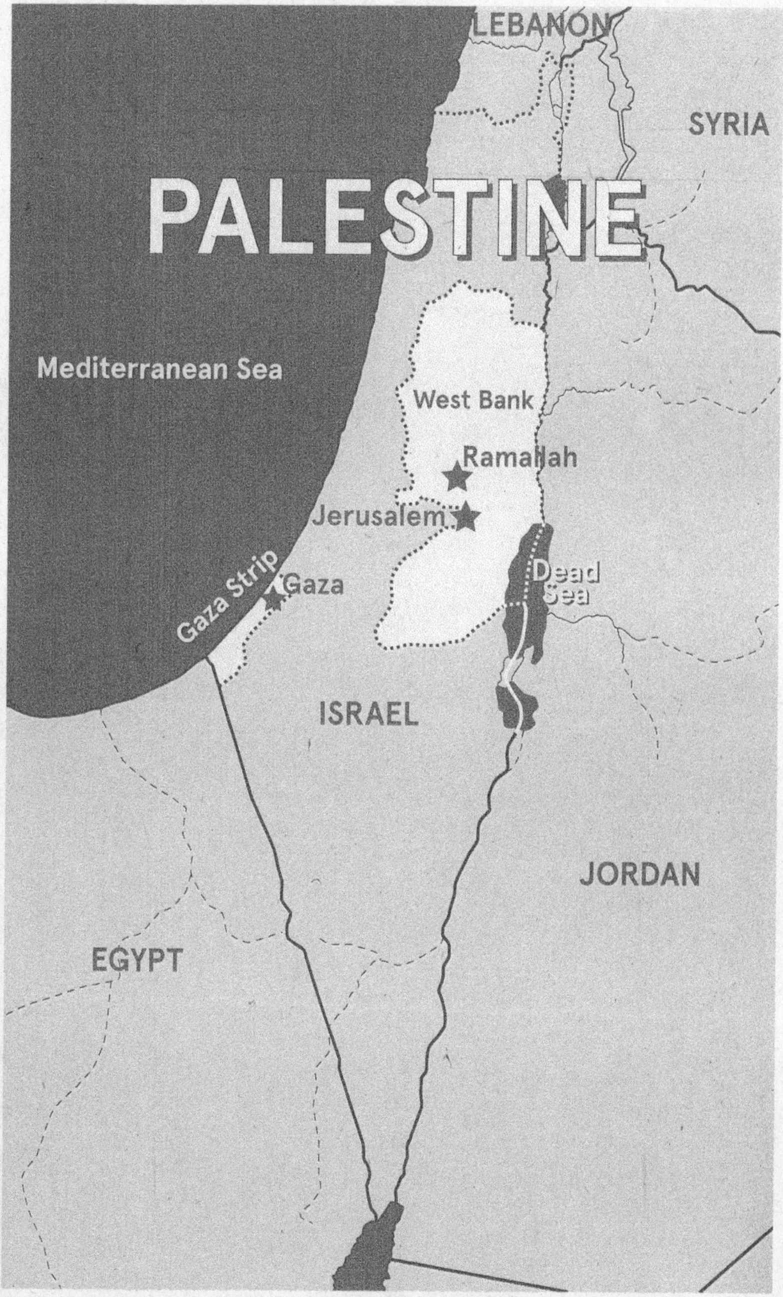

PALESTINE

LEBANON
SYRIA
Mediterranean Sea
West Bank
Ramallah
Jerusalem
Gaza Strip
Gaza
Dead Sea
ISRAEL
JORDAN
EGYPT

Evaporation

(Palestine)

How does the Union Jack make you really feel?

~~~

'I pass by these walls, the walls of Layla
And I kiss this wall and that wall
It's not Love of the walls
that has enraptured my heart
But of the One who dwells within them.'[74]

Not my interviewer having a moment, but the words of Qays Ibn Al-Mulawwah, a seventh-century Arabian poet who came from what would now be known as Saudi Arabia. In love with Layla al-Amiriya, who loved him back, he wrote poems to her which ensured he was labelled as majnun, mad, and so her dad said, no way are you marrying him, here's another, you'll be his wife instead because he and his tribe are powerful, they live next to the freshest water in this part of Arabia and, even in the seventh century, there was no greater power than water in a desert.
~~~

Their story became universal, most notably through the version of it written by famous Persian poet Nizami Ganjavi in the twelfth century. Lord Byron called it 'the *Romeo and Juliet* of the East', though it beats Shakespeare by about 700 years, so really *Romeo and Juliet* is the *Layla and Majnun* of the West.

I don't think the vetting officer had time for poetry, or traditional love stories. Though what is British exceptionalism but a love story, a national myth that must be repeated and retold, again and again? One that he wanted me to agree with, seemingly by rejecting any part of myself that could be traced outside its borders. And strangely, nothing else seems to elicit the warning bells announcing a mythical Great Britannia naysayer as much as public support for the rights of Palestine, specifically a return to the UN-brokered borders of 1967.

– *Do you support this too?*

– Yes.

– *Why?*

– Why wouldn't I? I studied International Politics, I wanted to work at the United Nations, why wouldn't I believe that their resolutions should be adhered to?

– *Surely there's an affinity too?*

– As a human?

– As an Arab.

~~~

Arab Palestine. Wall to wall with Laylas and Qays unable to be together because of water. War. Slaughter. Walls. One of the most famous walls in contemporary times is the West Bank Wall, built by Israel in 2002 during the Second Intifada, a Palestinian uprising against Israeli occupation. The uprising, which included attacks on Israeli cities, gave Israel a security justification for the wall, which is often described as a 'fence', although the vast majority of it is built with concrete breezeblocks, sometimes up to eight metres high and three metres thick. In 2004, the United Nations voted overwhelmingly in favour of the wall being removed (144 to 4), deeming it illegal under international law.

The wall has been criticised as a way for Israel to maintain a de facto border, appropriating Palestinian land to build it, with dehumanising results for the Palestinians, such as cutting family members off from each other, decreasing access to healthcare and severely curtailing their ability to work and travel around the West Bank area. In 2019, B'Tselem, the Israeli Information Centre for Human Rights in the Occupied Territories (meaning
~~~

West Bank and Gaza) lists 178 active checkpoints along it, most of which allow Palestinians through only on foot. Many checkpoints are open only a few hours a day for Palestinians and they need to gain permission papers from the Israeli government in order to pass through the main ones. There are also temporary roadblocks and pop-up checkpoints included in this list, cutting off thoroughfares and accessibility to necessities such as schools and hospitals, as well as one side of a village from the other.[75] Fences and checkpoints, checking . . .

***

Let's be the protectors of poetry
let's pull bricks down
with the tricks of words
and build them up again
when the sky no longer burns shadows
where we once spoke.

Let's spill dark ink on the sand
use the forgotten fingers of our hands
to fight with tides
that try to wash us away.

Let's mark walls with the blood of our tongues
to know our lives will be heard

even when they are reduced to nothing
but reddish grey rubble.

Let's displace sentences
until they are so at home
in every metal-doored house
nobody is able to tell where they came from any more.

Let's throw darts dipped in vowels
so when they pierce skin they bring not only pain
but the ability to begin at the beginning of language
again.

Let's put out our cigarettes on full stops
allow the ash-drowned letters
of doomed love to look through glassless windows
with shards of a hope made whole.

Let's stare at the sea
until similes sting our eyes
until we agree,
standing on a rock older than poetry,
that this land can only ever belong
to those who love.

No matter what the checkpoints say
or how loud the rockets scream
we know that love is enough

because there is nothing more
and we learnt that through words
when touch was disallowed
and so read me the poems as I drown,

my love.

***

Love. One of my firsts had soft lips and soft moments. But mostly he was deeply troubled and he searched for salvation through wraps of speed, televised football and the mouldability of my bones. If my cheek could bend just a little from one of his knuckles, perhaps he imagined he could imprint a version of life for himself that didn't revolve around drugs, prison, porn and Cash Converters. My skin a testing ground for how much power he could possess, despite statistics and predictions and the trajectory of every person he knew. Why I suffered being this testing ground for so long is beyond these pages, though the clues are scattered here and there.

***

Her arm just about fits through the letterbox, course it does, skinny little wrist like that. He always says if she'd got that job at Pizza Hut she wouldn't have been able to break into his house any more. She doesn't really like pizza, so disagrees, but keeps that to herself.

Her hand is flat, palm to the floor, as she pushes her arm in, but now is the hard bit, she has to turn it sideways, as if she's about to bring it to her other hand to do a huge clap. She twists her letterboxed hand round as far as it will go, anti-clockwise, ballerina fingers wait for the cold of metal to meet them and once it does, forearm burns and arteries start to get a little scratched against the tarnished copper of the letterbox frame, as she pushes her fingers down, unnaturally, to make the latch lower. Can't – Reach – Yet. She urges her arm in a couple of millimetres further and grimaces in a way she never would if he was watching and, finally, the lock clunks down with the touch of fingertips and the door creaks open.

His bedroom is immediately to the left of the front door. Or it can be entered from the kitchen. Two exits come in handy sometimes, and other times not so much. Perhaps it was supposed to be the dining room. Maybe it would be if his dad lived on his own like he wishes he did. The living room is directly opposite. It's stacked full of original punk rock and disco records and an actual jukebox. She would love to use the jukebox; its red lights and gold lettering seem to pulsate when she peeks at them through the keyhole of the door. She's not allowed in there on her own, because her boyfriend doesn't trust her to answer questions from his dad dishonestly enough. The dad is a big deal in the local National Front and her surname marks her out as 'other', even if her skin tone has so far allowed her a pass from this man who is so hardcore he bans Indian and Chinese takeaways from his sofa, fish'n'chips only round here, thank you very much. She never says, 'Well, that British staple came here with Jewish immigrants, you know?', and she never says, 'Why do you have so much hate in your heart yet so much good music in your head?' She's scared, she's a coward, she's a kid and she knows enough to admit the fear of being found out as foreign made her decline an invite to join them on a booze cruise to France in case he saw her passport.

His son is her boyfriend and before she stuck her arm through the letterbox, she knocked on the window of his bedroom and got no answer so she knows he's asleep, although she knew that without knocking. He never wakes before midday, but she always knocks anyway just in case he'd later say she was trying to spy on him and use that as a reason to lock her in the bathroom or something. And knocking isn't much of an effort, is it? It isn't.

She gently pushes open his door and he's unexpectedly sitting up in bed, football-team duvet covers over his knees, pull-out centrespreads of blonde women with balloon breasts covering the wallpaper that his dad chose, a bumpy cream variety that merges into strange, fantastical visions when her and him are high, like unicorns sneezing and cowboys in a shoot-out. She prefers the bumpy wallpaper despite its old-fashionedness, but he insists that the posters have to stay if she wants him to bother getting out of bed every day. Which she does of course, because there's always so much to be getting on with in the town centre or down the park, somebody's birthday or bought a new car day or it's a sunny day or a Friday. Today, the task at hand is opening her exam results. GCSEs! She was predicted good grades, but she wants great ones. She wants to get on in life, she wants her and her boyfriend who is sitting there staring at her with the meanest blue eyes she's ever seen to get dressed up for balls and awards ceremonies and she wants to buy him everything he's ever stolen. She wants them to have mini witches and wizards who live in a big house with a garden and even a swimming pool and a wooden playhouse maybe up a tree. It really could happen. She could be holding their golden ticket to transformation. This sealed brown envelope in her hand will reveal how valuable she is, if she will be welcome enough in a room of people who've never woken up wondering if they're still alive.

She doesn't ask why he's up early or why he's holding his own brown envelope in his hand. Hers is A4-sized, which trumps his, so she holds it up, excited, jumps on the bed with all her body, the weather too warm for her school jumper to still be on, but the balloon-breasted women on the walls make her self-conscious, so she keeps it as layered as possible until she's drunk enough to not care. He doesn't even say hello, which isn't too unusual, so she gets going, tells him about the anticipated opening of results, about her eleven GCSEs that she completed with relative ease considering she spent most of the year doing speed in living rooms with him and his mates, listening to 'I Don't Smoke da Reefa' blare from pirate radios and working in a kids' clothes shop to pay for both their partying. He tells her to go on then, why are you waiting for me? She says, because it's the start of our life, I want you to witness it. He says, yeh, yeh, witness this you stupid —.

Throws the small envelope that just moments ago she felt superior to, at her nose.

Open that.

She does. He already has but what he means is take out the paper inside, unfold it and read it.

Read it out.

She does.

Chlamydia, positive, all others negative. That's not that bad then is it?

Is it as bad as this?

The duvet's logo folds under his knees as he raises himself out of bed like Lazarus and grabs her shoulder hard, the envelope dropping to the carpet, her voice box limp now, lips turned down into a scared frown, waiting for the storm to pass before even thinking about making any noise except the slightest involuntary whimpering. Shouting makes things worse. Talking makes things worse. Crying makes things worse. Breathing must stay even. Keep eyes open. Limbs pliable. Do not struggle, struggle indicates a lack of love and a lack of love ignites worse. Worser. Worser should be a word. If girls had made up the lexicon it would be. Worser and worser. Her face now turned against that bumpy wallpaper. She's almost wishing for the posters with their balloon breasts, at least they won't leave imprints in her cheeks, at least they might provide some comfort, some distraction from the pain of jawbone against wall, hoping he won't break her nose this way, imagine the blood stain on the cream bumps, imagine what they'll become when they stare at them stoned one morning after raving. No, he won't go that far, he just needs to get it out, the anger, the despair of being abandoned. It's hard, to grow up like he has, she has to be patient, allow him to work through it, she is here.

You've been with other people; you've given me the clap.

Her self-enforced no talking when he's like this means she can't say that the clap is actually gonorrhea, you've just got chlamydia, which doesn't yet have its own nickname.

What's his name? Ay? You lost your tongue? Your dirty tongue that's –

His spit is on her face now, he is so close she can smell the strong tea with a splash of milk, any more than a splash is a conspiracy to give him allergies and cause his intestines to break down. She can smell a Findus pancake, cheese and ham, she thinks. Do not close eyes. He'll cry soon and then she can be Mary Theresa Florence Mary Margaret, stroke his back and say it's okay baby, it's okay now, I'm not mad don't worry, shhh now.

But for now, he presses against her, his forearm keeps her head to the wall, his thigh angled to trap hers, his free hand scrunches up the STD test results into a ball of icebergs and he pushes it onto the tip of her lips and she opens them all by herself and he wants her to eat it.

Eat it, eat your own dirt and –

It doesn't fit but she tries to stretch the mouth she was given to be the right size for this typed page of evidence and then – the phone rings. This doesn't always mean saved by the bell, so she doesn't get excited until he loosens his grip slightly, then altogether, then leaves, her own weight now pressing herself to the wall as he goes to the jukebox room to answer the phone. She hears the echo of laughter through the wall. It's a good call, then, that's good. She knows what to do now. She peels her face from the wallpaper, marked as she knew she'd be by the bumps his dad chose all those years ago when life seemed exciting, a new house, a new beginning. She takes the paper ball from her mouth, places it on a shelf next to empty champagne bottles and mugs with Union Jacks on. She doesn't want to throw it away, just in case. The chords of a song come through the door, Prince is on the jukebox, 'Party like it's 1999'. He bounds through from the kitchen, doing silly dance moves, his pelvis thrusting and his nose crinkled. She laughs. He picks up her A4 envelope from the carpet.

Come on then, let's see what you got.

You do it, I'd like you to.

He does. He scans the papers.

Well done baby. Can you make me a tea?

Hands them to her. Seven A*s, one A, three Bs, one C.

Yeh, course I can.

She breathes.

***

The Union Jack flag was always on his roof. This is not a metaphor, though it has grown to be one in my mind. The council had ordered the flag to come down. I think more than once they took it down by force. Each time, the defiant dad would put it up again, far prouder of his country than it was of him. The days when I would escape the boy's wrath from a window, climbing down long ladders propped by the side of the house, or drain-pipes if the ladders weren't there, the flag would flutter in the corner of my eye, jubilant or flaccid, depending on the temperament of the wind that day. So of course, a flag whose imagery is everywhere in such memories became associated with violence against my body and mind, but I didn't know that until the interrogator asked:

– *How did it make you feel, that he put you through all of that at such a young age and had a Union Jack flying proudly on the roof?*

– I don't know.

– It's not a trick question. Look at me. I'm not trying to trick you.

– I don't know how the flag made me feel, I don't –

– The flag in relation to what he did to you, Sabrina, that's what I am specifically asking here –

– I don't know what you're asking?

– Does the Union Jack . . . Does the British flag . . . Our national flag . . . Does the Union Jack . . . Does the British flag . . . Our national flag . . . The Union Jack . . . the British flag . . . Our national flag . . . is it inextricably linked with violence, to you?

~~~

Britain went into Jerusalem, in Palestine, in December 1917, accompanied by tanks and guns and the Balfour Declaration, composed by a British aristocrat who was the UK Foreign Secretary at the time, Arthur James Balfour, 1st Earl of Balfour. The Balfour Declaration said that the land of Palestine would now become a national home for the Jewish people in a country of people who were, in the vast majority, not Jewish.[76] Palestine was not Britain's to give away to anybody else, no matter how in need anybody else might have been. Of course, genuine support of Zionist endeavours was a very small part of the reason the British did what they did. Britain needed –
~~~

wanted – north-east access to the Suez Canal, which Palestine could provide better than any other location. If British Zionists were to be put in control of the area, the economic and military benefits to an empire increasingly dependent on the Middle East for its survival would be obvious. To that end, Britain also wanted a base next to Egypt, as the country's campaign for independence from the Ottomans was gaining rapid progress (it happened in 1922) and Britain wanted to capitalise on the fall of Ottoman territories as much as possible, especially ones as important as Egypt. Britain also wanted to win the First World War. The Balfour Declaration was officially approved in October 1917 following a UK Cabinet meeting discussing how such a statement may elicit support for the Allied efforts from the worldwide Jewish population, and particularly from the United States. It followed a six-month stalemate in Beersheba, southern Palestine, a stand-off between the British Empire's Egyptian Expeditionary Force (EEF), an imperial military formation, and the Ottoman Army and its German allies. Coincidentally, on the same day the Declaration was approved, the Battle of Beersheba began, ending the stalemate and resulting in the retreat of the Ottomans and their allies a few days later, avoiding the Ottoman/German capture of the Suez Canal, which would have perhaps been a fatal blow to overall victory in the war and to the survival of the British Empire. The Declaration was the ideological formalisation of justifying an occupation of Palestine, whilst the Battle of Beersheba enabled the physical British occupation to begin. The Allied forces broke the Ottoman line at Gaza, and continued through to Jerusalem, where they ruled under a League of Nations-approved mandate until 1948, when an independent State of Israel was declared by Jewish leader David Ben-Gurion, with backing from the United States.

Palestine has been subjected to explicit military, emotional, physical and psychological violence ever since. In its World Report 2021, Human Rights Watch notes that 'Israeli authorities in 2020 systematically repressed and discriminated against Palestinians in ways that far exceeded the security justifications they often provided'.[77] The report details how 2 million Palestinians are denied their right to freedom of movement in Gaza, and by purposefully limiting their access to water and electricity, as well as harsh restrictions on the exit and entry of goods, the Israeli authorities have forced 80 per cent of the population to rely on humanitarian aid. In the same year, the Israeli government committed a war crime by transferring additional Israeli citizens into settlements in the occupied West Bank, demolishing 568 Palestinian homes and structures in the process. This was the highest rate for four years and the authorities have declared they will keep flouting international law to build more settlements in the West Bank. This action is illegal due to many resolutions voted on and agreements made by the UN, the most recent being in 2016, when fourteen member states, including the UK, but not the US, voted that Israel's settlements in any Palestinian territory that had been internationally formalised in 1967 had no legal validity.[78] Tensions continue to escalate globally around this issue, with a two-state solution being favoured by most political leaders since its emergence as an ideal in the 1993 Oslo Peace Accords. This solution is unviable for most ordinary people due to the expansion of the Israeli settlements and the changing power dynamics of the region, which have seen Bahrain and the UAE sign a Trump-pushed 'Peace Pact' with Israel, as Iran and Turkey become the neighbours they are arguably more threatened by. Considering all of this, Britain may not be as directly involved any more, but the water access it fought and lied for still plays an integral part in maintaining this violence against the indigenous population.

~~~
~~~

Gaza is one of the most water-stressed places in the world. And yet there is so much water right by it. At the time of writing, Gaza's water crisis is predicted to soon make it completely uninhabitable, with at least 500,000 children alone having restricted access to clean and safe drinking water.[79] The area has water. The Wadi Gaza is considered one of the most important wetlands in the Eastern Mediterranean.[80] But it is overrun with raw sewage, with around 30,000 cubic metres flowing from it into the sea every day. Because of the Israeli blockades on Gaza, including a ban on the entry of construction materials, the delivery of cement, pipes and equipment being either delayed by years or never allowed in, there are very few possibilities for maintenance and renewal of destroyed sewage and water management systems.[81]

This inability to repair and rebuild sewage treatment facilities means that the Mediterranean sea off the 45-kilometre coast of Gaza contains toxic Escherichia and Enterococci bacteria, both of which originate in human faeces, at levels far above World Health Organization standards. At the UN Human Rights Council in October 2021, the Global Institute for Water, Environment and Health and the Euro-Mediterranean Human Rights Monitor made the following joint statement: 'The long-term Israeli blockade has caused a serious deterioration of water security in Gaza, making 97% of the water contaminated . . . The residents of the besieged enclave are forced to witness the slow poisoning of their children and loved ones.'[82]

The region's coastal aquifer is a rainwater-fed resource that historically has enabled agricultural civilisations to flourish, from the Romans to Theodore Herzl's Jewish state. Although millions still rely on this aquifer, from Haifa in the north-west of Israel to the Sinai border town of Al-Arish in Egypt, this shared Israeli, Egyptian and Palestinian resource is becoming poisoned by groundwater salination caused by over-pumping and by Gaza's cesspools and sewage. Egypt and Israel have the resources to find alternatives; Gaza does not and is actively stopped from finding them by Israeli blockades and sanctions.

This includes Israel's 'systematic targeting of electricity networks and ban on the entry of fuel', according to the Al Mezan Centre for Human Rights.[83] Without stable or reliable electricity, it is impossible to maintain water wells and sewage treatment facilities. Israel also denies Gaza access to concrete, basic construction and technological maintenance supplies. Israel bans the entry of steel pipes larger than 1.5 inches in diameter into Gaza, with the knowledge that desalination and water treatment plants require pipes with a diameter between 2–10 inches.[84]

At the October 2021 UNHRC session, officials estimated that 80 per cent of Gaza's untreated sewage flows directly into the Mediterranean, hence the toxicity levels present in the water on the coast. They went on to state that this humanitarian catastrophe causes around 25 per cent of all diseases spread in Gaza and 12 per cent of the deaths of young children, who develop intestinal infections due to contaminated water. The organisations agreed that the situation was greatly exacerbated during the 11-day Israeli offensive on Gaza in May 2021, when 290 water supplies were damaged or destroyed in the attacks, including 3.7 miles of water pipelines.[85] Sewage networks were also obliterated as part of the offensive. These networks were already severely weakened by the Israeli invasion of 2008, where $60 million worth of damage was caused, and repairs were mostly stalled due to the blockades.[86] The May 2021 offensive included targeting and destroying the only desalination plant in Northern Gaza. Al Jazeera reported in October 2021 that locals were desalinating water themselves to sell to a desperate population, and that the municipal water supply was so toxic it killed vegetation.[87]

In February 2022, the United Nations Development Programme put together a $66 million plan to 'save' the wetland territories of the Gaza Valley[88] and international non-profits are supplying clean water tanks that a minority of the population can access,[89] but what will these efforts mean in a besieged state of more than 2 million people?

And it's not just Gaza. Ramallah, the administrative capital of Palestine whilst it is not allowed an actual capital (which would be East Jerusalem, in territory that the United Nations, even the United Kingdom, but not the United States of America, currently recognises as occupied land), has more rainfall than London. Israel takes 80 per cent of the water from the West Bank aquifer and yet is constantly written about in the Western media as a country which has 'solved its water scarcity problem with technology and tenacity'.[90]

Over seven months during 2011 Israel systematically destroyed eighty-nine water-related structures in the West Bank including cisterns and wells.[91] Demolishing a well so people cannot reach water that has fallen from the sky to sustain life. Death in exchange for the draining of all a place has to give. Why? What is the point? When I visited the West Bank in 2014, as an author touring with PalFest, the Palestine Festival of Literature, I could certainly see no answer except the usual ones – power, greed, pride.

***

In Hebron, once-heaving streets of silent shops shadow our path with shuttered rust, we walk quiet, throats full of unanswered bulldozers. Soldiers block off neighbourhoods to those people whose ancestors' bones have carbonised the ground. Children playing chicken over machine-gun motorways the size of two pairs of khaki-covered conscripted teenage thighs. The price for their game might be a sigh or a slap or a shout or perhaps one day a bloom from the stem of the gun held by hands that haven't yet learnt their lines. We scatter ourselves around on Tarmac two shades darker than the overcast sky, lean against concrete blocks to take in the scaffolding of dismantled existence. A settler approaches with a video camera (the kind that used to be called handy before they actually were); we ask if he's filming because he appreciates our dress sense or if he's making a documentary about the few who come to Hebron to witness the apartheid, tell of it what they can to the outside, and he says, with spit spinning around the wheels of his words, 'I'm filming you for god.'

In the middle of that now-sunlit midnight street I began to daydream about god watching our group of writers on a flat-screen TV – once the settler sends it via courier or however they do this kind of thing. I wonder if god would have HD or even 3D? Would this supreme being need those rubbish blue and red cardboard glasses to see that in? Would god have a remote or would it all work via mind control? Would the back of the TV get dusty all the way up there? Would god watch it alone? With popcorn? Clean water? With pick'n'mix sweets that send torrential downpours of eaten-too-much-too-quick sick down to the ground a few hours after, the world having no idea what caused these strangely coloured lumps to tumble down one morning?

Then I was wondering what else would be included on this exclusive home video for god. Would it show certain settlers filling plastic bags with bleach, throwing it down onto the Arab marketplace to make clothes unsellable? Or how about a little cameo from the settler school built on top of the Palestinian one, literally crushing its core, which now requires a one-and-a-half-hour detoured walk to get to and from every day, since the army blocked the entrance alleyway that used to get the local kids there in ten minutes. I'm sure that would make for scintillating TV.

But the really juicy stuff, I assume the filmmaker settler who directs movies for god will leave until last.

Close-ups on the roads Palestinians can't drive on but are certainly encouraged to die on, ambulances included in the ban so stretchers held in hands that have outlived their lives rush people to what care they can get. Sunken canvas pulsing muscles doing what wheels were invented for.

Perhaps a fitting finale would have to be the dehydrated six-year-olds arrested for allegedly throwing pebbles against the thunder of Israeli Defence Force patrols. Held under military law until fines take them home to a place that's only allowed a lock so no light can get in, such permission only granted to Hebron's windowless walls, choose light or a lock, another psychological weapon of woe. When the children turn twelve, suddenly those pebbles they touch can become army slabs of rock they will be tied to, pecked like Prometheus but before they grow old enough to give fiery gifts. And the credits for this movie – who would be on it? A disturbing amount of worldwide names, mainly American and European. It's doubtful god would be able to watch all the credits roll, as being god requires strict time management, I imagine.

In the end, we turned our own cameras on the settler saying: 'We'll be sending this to you once god's had a preview. You won't be invited to the cinema screening, but we have a feeling you'll steal someone else's seat anyway. Good day.'

***

I once watched the mesmerising work of the Palestinian musician and performer Dina 'Amr, who performs under the name Bint-Mbarah. In this particular piece, *time flows in all directions_ water flows through me*, 'Amr collected the songs and words of the last generation of Palestinian women who perform traditional Sufi rain-summoning songs, interspersing their recorded vocals with her own live ones. Sufi rituals were part of many localised traditions that were lost when Palestinians were forced from their land following Britain's actions. Significantly for what much of this chapter has looked at, 'Amr insists that when she performs the rain-summoning pieces, she is genuinely trying to make it rain, especially when performed in places such as Amman, which were then experiencing a drought. By summoning rain using these dormant Palestinian Sufi rituals, she is reclaiming culture, but also the water that has been stolen from the Palestinians.[92] An act of love, through water, two things we should all have, but that seem increasingly difficult to guarantee.

***

I stood in a garden and I told it all I knew about us. It listened in the same way that you do. The larger leaves, distracted by the wind, made attention-deficient shadows on the small patches of grass, and the creased petals of wild flowers peeked hopefully through windows bordered with thorns. The dry, loose mud seemed the most interested, rolled closer to my voice by sympathetic insects who know all too well the difficulty of not sharing daily breath with the one you love. Some nettles nodded at my monologue, tentatively at first and then with absolute urgency as they noticed my eyes might drip; it hadn't rained in a week.

The tree in the heart of the garden shared its stance with you, somehow there was your arm in the branch that bounced aggressively as my words got louder. An olive, disturbed by the bounce, bullet-fell to hit the roof of the shed, its wooden tiles torn with pencil shavings of neglect. The olive's journey lent its rhythm to my rant until it reached the murky, malnourished pond where it slipped in with an ancient ripple and I did a subtle throat-clearing as I felt they were all getting a bit too distracted now. I stood on the two planks of the bench that weren't rotted and I think that the height helped what happened next. A broken piece of terracotta pot seemed engrossed in the way my lips trembled when I said your name. And so I said it again and again and again and that's when the rains came. Heavy, gutter-filling drops. My audience drank and came alive, so alive now. They had such a lot to catch up on they didn't even notice how I drowned, my open mouth bubbling, a stone water feature.

***

– ***I'll ask the question again, Sabrina. Our national flag, is it inextricably linked to violence, to you?***
– Yes, sir. I think perhaps it is.

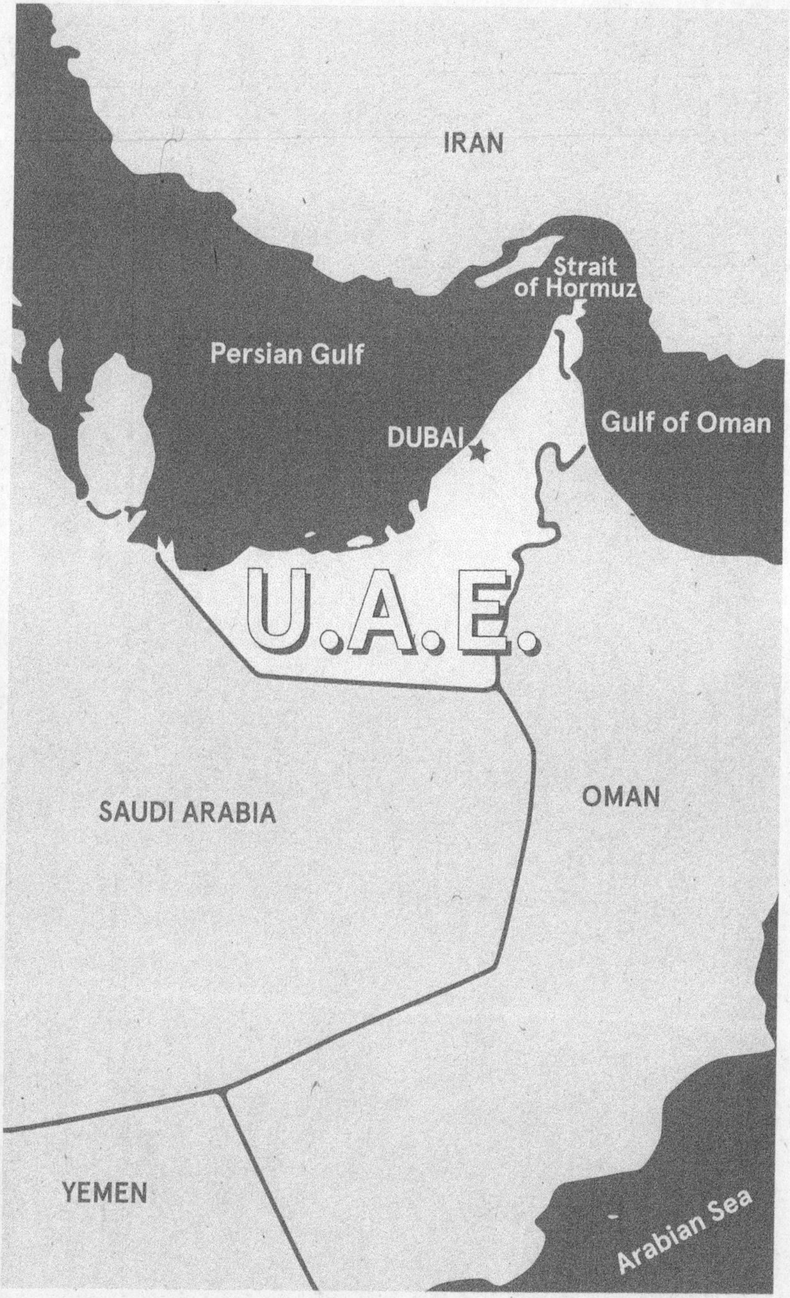

IRAN

Strait
of Hormuz

Persian Gulf

DUBAI

Gulf of Oman

U.A.E.

SAUDI ARABIA

OMAN

YEMEN

Arabian Sea

Permeable

(United Arab Emirates)

Why Dubai?

~~~

After a particularly long night during the MOD training programme, where I had slept with a decommissioned SA80 machine gun in my sleeping bag, covered by a branch-tied tarpaulin, and woken up to a slug slithering from the leaves beneath me to my cheek, I decided to take the consequences and leave the programme for a week. Perhaps I mitigated those consequences by saying I was sick. I can't remember. I only remember booking a flight to Dubai to try to find my sort-of boyfriend at the time, who was there 'on business' and had been unreachable throughout the last month. Dubai, at that time, was the Costa del Sol for Brits from the Global Majority, with the added complexity of being somewhere that still felt like a British protectorate in many ways, with English the default language of business and leisure and British companies present at every turn. Here in Dubai, British accents and passports elevated people to positions of
~~~

power where dubious moneymaking was without much consequence – a privilege many white Brits did not need to travel for. That lasting illusion of British colonialism was a result of a long relationship between the UAE and Britain. Prior to its independence in 1971, the UAE was part of the Trucial States, a collection of sheikhdoms that stretched from the Straits of Hormuz to the west coast of the Persian Gulf – 32,000 square miles of loosely defined tribal groups, which would include present-day Qatar and Bahrain. Pirates had worked in the seas of the Trucial States for centuries, and when Britain began attacking them to defend their ships bound to and from India it prompted ties to be made with the various sheikhs, who also wanted to rid the area of pirates. All the sheikhs of the Trucial States agreed to a truce between themselves, brokered by Britain, and pledged not to cede any land, nor make any treaties, nor settle any dispute, with anyone but Britain. Without an official army themselves, their territories were also threatened by pirates, so it must have seemed sensible to acquire the protection of the largest navy in the world. Perhaps nobody imagined it would be over 150 years before they could leave Britain's enforced 'protection'. Though independence eventually happened, with Bahrain and Qatar becoming their own states and Abu Dhabi, Dubai, Ajman, Al Fujayrah, Sharjah and

Quwayn joining to become the UAE. Before it took down its flags, Britain explicitly allowed Iran to occupy three islands in the Strait of Hormuz that Iran alleged Britain had wrongfully granted to the Trucial States during their time as a protectorate, and that they wanted back to protect Iranian oil tankers in that body of water. Britain's official reason for not stopping the occupation was to save the UAE from becoming embroiled in a military dispute on its first day of independence. Iran still controls these islands and the dispute hangs over their relations with the UAE today.

After all of that, Britain has never really left. Even though the estimated 200,000 British migrants living in the UAE is a small percentage of the more than 8 million expatriate residents, there are over 5,000 British companies currently operating in the UAE. For comparison, Britain is Hungary's twelfth most important trading partner; it is a country of a similar size to the UAE, but has only 900 British companies operating there.

I don't know what business my sort-of boyfriend was doing in the Emirates – if I did, I wouldn't write it down. Our relationship was mainly made up of lingerie, nightclubs and shadows, so it may be that during our years together I never knew what he was doing, unless he was in my bed. Regardless, I was in my early twenties and I wanted to let him know he could not be uncontactable for weeks, no matter what the scenario. I needed someone to talk to after a night curled up with a machine gun, after all. I also wanted an adventure of my own making, even if the impetus was the unoriginal one of chasing a man. I likely framed it as saving a man; even less original. I had realised during the MOD training that me and state-sanctioned adventures were not going to be compatible, so I had to create my own. Dubai was the perfect Lego board for doing so. I read books about marketing on the plane and, when I arrived, set up meetings all over the city as the Creative Director of my own PR company, created on the flight. It was frivolous and fascinating and fun, in the way colonisers of the past might have found their new destinations. I was right there with them, those figures of history I lacked any sympathy for, avoiding my own privileged, patriarchal role in the world as a British person, just as they had. I ignored the minibuses of indentured labourers being carted off to keep building the city higher and higher before they'd get their passports back, if they survived long enough. I met up with old uni friends in beautiful cafés, mostly Arab women who had found comfort in Dubai's relative freedom, and professional success due to its commitment to becoming a global centre of all industries. Its ability to have a future depends quite literally on its ability to become a world leader in water management and conservation, meaning many of the women I knew had well-paid junior roles in this area and are now in senior roles across the world, in part thanks to the UAE's foresight in promoting STEM (science, technology, engineering and mathematics) training specifically for women. Some of them have stayed in Dubai, where they are much needed.

The population of the city has more than doubled in the last ten years. A decade that has been the deadliest for Middle Easterners due to armed-conflict fatalities[93] and the driest for at least 900 years, certainly the driest decade on record. Lack of rainfall can be counted as one of thousands of reasons that the conflict in Syria happened in the way that it did, with some sources stating the years preceding the start of the conflict saw the worst drought for 500 years, resulting in crop failures, increased food prices and a mass migration of millions of farming families to the cities, where resources were already stretched. After a decade of conflict, an estimated 15 million Syrians now lack access to safe, clean water.[94] Whilst it may seem ironic that in this decade of drought and destruction wealthy displaced Arabs from Syria, Lebanon, Egypt, Iraq and Kuwait have sought a new home in one of the driest places on earth, the Arabian desert, that has happened, and it is part of the reason for this population – and economic – boom. Water conservation, renewal and management are absolutely key to the UAE remaining a destination of choice for the global elite. Its Water Security Strategy 2036 is one of the world's most ambitious, utilising high-tech processes such as reverse-osmosis filtration (RO), also known as hyperfiltration, which is currently the best means of desalination, producing fresh water by removing salt from saline water. The technique has evolved in part due to research on how to remove the human body's impurities with dialysis machines and has created spin-off applications in filtering water in salt-water fish tanks. Using electric power, water pumps apply hydrostatic pressure to the saline feedwater, forcing it to overcome the osmotic pressure equilibrium and be driven through a semi-permeable polymeric membrane. Salt-water purifiers reject salt particles, allowing only fresh water to pass through, the flow of which continually cleans the membrane under a process known as crossflow. In order to maximise pressure, desalination plants use spiral-wound membranes mounted in high-pressure containers to maximise volume and surface area. In between each curve in the spiral is a mesh-separator that allows fresh water to flow out and sweep the membrane clear of particulate. The Taweela Al

Power and Desalination Plant in Abu Dhabi is expected to produce some 910,000 cubic metres a day, making it the world's largest seawater reverse-osmosis desalination plant when it becomes operational in 2022. Abu Dhabi has already completed the world's current largest desalinated water reserve with the Liwa aquifer, with a capacity of 26 million cubic metres.

Desalination is effective but environmentally destructive, mostly due to the resulting brine increasing the seawater temperature, salinity, water current and turbidity. This in turn harms the marine environment, causing fish to migrate, increasing the presence of algae, nematodes and tiny molluscs. So reducing water consumption by at least 21 per cent is also one of the UAE's main objectives, despite its continuing population growth.

Dubai is recognised as one of the fastest-growing cities in the world. Its annual gross domestic production has reached about $20 billion, with a relatively small population of 2 million people. The government strategy of shifting Dubai into a tourist and business destination and away from a reliance on oil production has evidently worked, despite its long-term viability remaining fragile. Dubai's rapid economic growth can be partly attributed to the free-trade areas and investment from overseas that the government has encouraged. Despite this growth and commitment to water security, rapid urbanisation has led to complex ecological, social and infrastructural issues. Dubai's only waste-water treatment plant, at Al-Aweer, is operating over its capacity (equivalent to 4 billion bottles of water a day), producing treated sewage which is far below the expected quality standards of the country. The plant, which was designed to treat 260,000 cubic metres of sewage a day, now treats 460,000 cubic metres a day, with about 3,000 trucks each day depositing 100,000 cubic metres of raw sewage at the Al-Aweer plant for treatment.

The world's tallest building, the Burj Khalifa, uses the equivalent of twenty Olympic-size pools of water a day just to keep it cool. Cooling and desalination also require huge amounts of energy usage, which adds to their carbon footprint, which ultimately leads to further environmental damage. However, even with these demands, the UAE uses less energy per capita than the United States and far less than other countries in the region, such as Bahrain, which is the third-largest energy user per capita in the world, after Iceland and Norway.[95] Dubai Electricity and Water Authority hopes to improve this even further via its carbon footprint campaign. If electricity can be made sustainably, then water does not need to be used to generate it, as solar and wind power don't require the application of water, unlike traditional power-generation plants, which are highly water-intensive. Data isn't available for the UAE, but, as an example, 40 per cent of the fresh water drawn in the USA is for usage in traditional power plants.[96] The thrust of the Dubai Electricity and Water Authority's Green Energy Push programme is to transform 30,000 buildings in the city into sustainable structures and to instal energy-saving streetlights.

Abu Dhabi is also hugely dependent on desalinated seawater, and in order to overcome this dependence it has developed a groundwater-monitoring system to prevent up to 40 per cent of water leaks in parks, pumping and irrigation. This system monitors water pressure and flow to prevent water wastage and detects network errors and technical problems.

Super-rich Emirati citizens with private companies have their own wildly ambitious ideas to solve the problem, such as towing icebergs from Antarctica to Fujairah in the hope of providing fresh water for a number of years, as well as adjusting the region's climate by the icebergs' attracting clouds and rainfall.[97] Around 70 per cent of the earth's fresh water is stored in glaciers, which build for tens of thousands of years until chunks break off and form icebergs, which in turn can be monumental in size and, consequently, fresh-water capacity. The tallest-known iceberg in the North Atlantic reached 168 metres above sea level. Since the bulk of an iceberg is below the water, the entire berg was estimated to be as tall as a fifty-five-storey building. Iceberg B-15, which calved from the Ross Ice Shelf of Antarctica in 2000, was half a mile thick and covered an area of about 4,500 square miles (about the size of Connecticut). A far smaller iceberg, one that measured around 3,000 × 1,500 × 600 feet, would contain around 20 billion gallons of fresh water. Icebergs only last around three to six years, so eventually this fresh water will become seawater, if it is not harvested as fresh water. These are the facts that supporters of the idea use to dispel concerns that 'playing God' with the environment in such a way could cause unimaginable ecological disaster.

The outcomes from all these plans to manipulate Mother Earth on the future viability of the UAE remain to be seen. But back when I was visiting, nobody was talking about icebergs. My own days there go something like:

***

– Dubai, I'm in fucking Dubai babe. Where are you? Better answer your phone yeh. I left training camp to come check you, haven't heard from you for like, two months. You're long babe. I been kipping in a sleeping bag under Brecon trees with an SA80 machine gun between my legs and my Ericsson by my head waiting for you to ring, to text, to anything – I thought you might be, you know, dead or whatever.

I stand in the Arabian desert waiting for my South London gangster, drinking the last of my last plastic bottle of water.

My skin drips though I stand in the shade of the hotel's polished gold entrance and finally, the motorcade arrives. Moody-faced boys with hard London accents. I feel safe with them in a way I can't explain except to say they are home.

– Get in then babes. You shouldn't have come though. I told you innit, man's got business.

– Yeh, so do I.

We lie in the ice of the AC and I face his lips, Red Bull-laced and missed. Proper missed.

– This training so far babe, it's took the piss a bit, but it hasn't put me off. I'm gonna go through with it. The whole application. The Developed Vetting for Top Secret Clearance. For Defence Intelligence. I won't mention you, obviously. But if I get it, me and you. I dunno.

– Course you'll get it babes, you're the Egyptian Guyanese Nancy Drew, that's like terrorism and drug kingdoms done really innit.

– Dickhead.

– Nah serious, I ain't gonna be a good look for all that, course not. They can't have someone working for them who gets with someone like me. You need a suited-and-booted boy. Man. Anyone you want, but not me. I know how hard you worked for this, miss. You take that shit. You deserve it. It's good out here for me. I can do big things, I can feel it. People got mad respect when they hear you're English and I ain't never even thought of myself as English before I got here, crazy innit.

– Yeh. Yeh, it is. So what . . . that's it?

It is.

***

I swim in the pool because the sea can't be reached without scorching the soles of my feet and I need water water water over my face to stop it being drenched by my silly little heart.

I can't make the rest of the meetings that week. Pretending to be in PR, testing if a life in Dubai was for me. I think it was only for him. And he's gone.

I meet a mate I went to SOAS with. Amalah. She's Emirati, an entrepreneur, trying to conquer the 'dating site' market in a place where many people pretend they don't 'date' before marriage, and apps at this time are generally considered a waste of phone memory, the universal comfort of their future necessity unimaginable. She brings me my first smile in days.

***

– The weather here is not conducive to taking lovers, ya Sabrina. It is so hot outside you sweat before you blink, it is so cold inside you don't want to take a single piece of clothing off. Most of the sex I have is in the shower. The government are desperate to save water. Only for those who live here, of course, tourists can pour shiploads of it into porous rock for pure entertainment as long as they keep sharing pics with their sun-starved friends.

So they've suggested this thing. Singing in the shower. As soon as you turn on the water, you start to sing your favourite song. Once it's done, you turn it off. Waste not want not. So, of course, I launched my own idea, Shagging in the Shower. Undercover, looks like a site that sits in Turkey cos they're slightly more relaxed about all that stuff, but, actually, anyone who knows, knows. We are growing every day. Expats and Emiratis like me, rated by their shower-shagging ability. And so we have this unbelievable USP. Environmentally friendly, minimum time investment, no need to worry about an outfit or sweatmarks or anyone catching you. People are renovating their bathrooms just so they can get better matches.

It's making a real change in how millennials view their water usage generally. We use so much more of it than we have. It's a crisis, really. Imagine, the British colonising a place without a river! But we did get bonus points for the tip of a land point that pokes into Iran AND controls sea access from the Persian Gulf to the Gulf of Oman to the Arabian Sea straight to India. Not even SOAS could've taught me that Imperial Britain's dependence on water resources in the Middle East would decades later make me the most popular dating entrepreneur in the UAE!

Anyway, how are you, ***habibti***?

***

I am heartbroken and soon back in Whitehall. The interview room, the Developed Vetting Officer – my interrogator. A teacup. Empty. Almost. No toast. Tea and toast is the cosiest thing I can think of. This teacup on this table in this room is not cosy. This cup of almost gone Rosie Lee is not making me feel relaxed. Beige pool at the bottom of the cup, I imagine it rising up, rushing in a tepid wave to cascade itself over the lapels of the mac, the colour of which it matches perfectly, the mac being worn by the man, the man who is demanding I answer his completely normal questions such as:

– *Do you have you a boyfriend? A man in your life who you are . . . intimate with?*

– What about a woman?

– *Fine, do you? Have a woman? Someone, whoever, you have sex with?*

– Nobody.

– *You are not having sex?*

– No, that's right, I'm not.

– *Would you consider yourself asexual then?*

– Would you consider that desirable for clearance?

– *That's not what this is about. This is about risk.*

– Is sex a risk?

– *It can be, of course.*

– Is non-sex also a risk then?

– *It can be, depending on why there is no sex.*

– Isn't everything a risk depending on the why?

– *Your British passport, the photocopied pages I have here, I was surprised to see so many entry stamps for Dubai. Four trips in less than two years.*

– I have friends there, from here, from Egypt. It's fun.

– South London criminal gangs enjoy it there at the moment too, apparently.

– Do they?

– The Muslim ones, anyway.

– Most of the British people I see in Dubai are from Essex.

– You are from South London though, aren't you?

– Yes. What can I say, the high-rise estates gave me a taste for the five-star-tower-hotel life. What has Dubai got to do with sex?

– For an attractive, intelligent twenty-four-year-old woman to not have any intimate relations of any kind strikes me as –

– PTSD? Necessity?

– A lie. Or otherwise, a kind of asexuality. Both carry risk, but if it is the first one, then I must ask, why would you lie? All your friends and colleagues that we interviewed corroborate that you have no love life, but they would know to lie if it was something that really needed hiding. Or maybe you hide it from them too? Such efforts would only be worthwhile for love, not for something as simple as a left-field sexual fetish for example. Who would need to hide something as basic and essential as love, I am genuinely intrigued here, Ms Mahfouz – why? Why hide love? Who is it you love?

What do they do? For you? For us? Who? Why? Who?
Why? Who? Why?
Who why who
why who
why
who
why

I dream of him every night after the last trip to Dubai.

***

The ghost that drinks tea.
He said:
'Princess, why are you sleeping with the light on?'
I said:
'I thought it might mean I wouldn't be able to see you.'
I put the kettle on and we spoke of how windy
it had been that day for this time of the year and all.
He chose a biscuit with the most chocolate on
and dipped it slowly into his strong tea,
the colour of a stormy beach.
I didn't have much to say –
words seem to fade into the air when there's a ghost
sitting right there in your kitchen.
But I was unable to look away from him,

his skin seemed to glow and my fingertips ached
to touch it, only to find something like cold cotton wool.
I shivered and he tutted;
*'I thought I made you feel all warm inside,
least that's what you said in my birthday card.'*
That was a while ago, I thought
but just smiled in answer and turned the heating up.

***

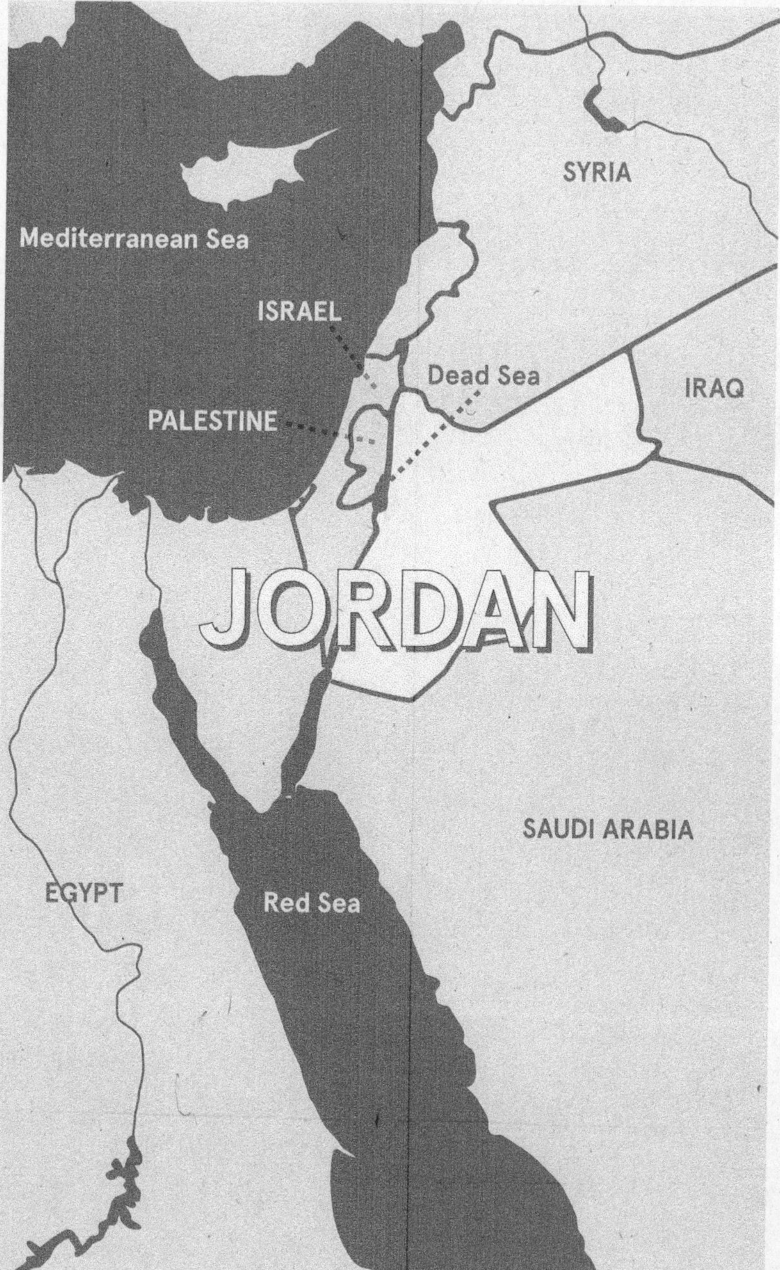
SYRIA
Mediterranean Sea
ISRAEL
Dead Sea
IRAQ
PALESTINE
JORDAN
SAUDI ARABIA
EGYPT
Red Sea

Confluence

(Jordan)

Where do you go to feel safe?

~~~

> All territory lying to the east of a line drawn from a point two miles west of the town of Akaba on the Gulf of that name up the centre of the Wady Araba, Dead Sea and River Jordan to its junction with the River Yarmouk: thence up the centre of that river to the Syrian frontier.[98]

The League of Nations stamped its approval on the above as the borders of Transjordan in 1921, an area to be indirectly ruled by Britain until 1946.

As the First World War ended in 1918, so did the Ottoman Empire, which had sided with Germany. Many of its territories, such as Egypt, were already being informally run by European powers after its protracted decline. Now it was officially over, France and Britain wanted to move quickly to cement their control of the old Ottoman territories and stop the power voids being filled by Russia
~~~

or the people of the region themselves – unthinkable! Revolt was rising against the British and French across the Middle East at this time. People did not want to swap one imperial power for another, especially as many had fought in the Great Arab Revolt of 1916–18, which had played a significant part in the First World War's outcome, helping to expel the Ottomans and create a unified Arab state from Aleppo in Syria to Aden in Yemen, which the British had promised to recognise. Now the Ottomans were no longer a threat, Britain did not plan to keep that promise. Along with France, which also wanted control in the area, the British knew that arbitrary borders with differing power centres, as well as a significant military presence, would help to obstruct revolt. In Transjordan, they appointed Emir Abdullah to ostensibly rule, whilst one of his brothers was made King of Iraq and another King of Syria. Any uprising against their rule was met with military force. Emir Abdullah assisted the British and French for decades, including by creating the Arab Legion, a reserve force which assisted in the European occupations of Iraq and Syria. He was able to negotiate Jordan's independence from Britain in 1946, at which point he became King Abdullah of the newly named Jordan. King Abdullah was part of the Hashemite family, descendants of the Prophet Muhammad who had ruled the city of Mecca in Saudi Arabia since the tenth century, but had lost their power to the British-backed Ibn Sauds after years of dispute and distrust.

Abdullah's father, Hussein Bin Ali, had led the Great Arab Revolt against the Ottomans from 1916 to 1918. The revolt had been based on the McMahon–Hussein Correspondence, a series of letters exchanged between Hussein Bin Ali and a British officer, Lieutenant Colonel Sir Henry McMahon, during the First World War. The ten letters were written between July 1915 and March 1916 and outlined how the British needed the Arabs to expel the Ottomans from the region, as they had declared war on the Allies. Britain wanted to protect its regional interests and avoid losing the support of India's 70 million Muslims by explicitly fighting against a Muslim empire. If the campaign was led by the Arabs, then it would benefit everyone, as in exchange for their leadership Britain would recognise pan-Arab independence once the war was over. The Arabs were militarily victorious, forcing the Ottomans from much of the region. The British, however, reneged on their side of the deal, particularly with the Sykes–Picot Agreement and the Balfour Declaration on Palestine. When Hussein refused to officially accept the British Mandate for Palestine, the British decided to give their support as rulers of the region to their Central Arabian ally, the Ibn Saud family, who proceeded to conquer Hussein's kingdom in the Second Saudi–Hashemite War of 1924–5.

Perhaps this partly explains why his sons, especially Abdullah, were so desperate to re-expand their Hashemite territories by complying with the British, despite this causing widespread distrust of him on the part of other Arab leaders and local populations.[99] He was assassinated in Jerusalem in 1951 by a Palestinian man from a prominent family. Abdullah was there to attend meetings with the director of Israel's newly founded intelligence agency, Mossad. A few days earlier Lebanon's Prime Minister had also been assassinated, supposedly due to rumours of a secret peace deal between Lebanon, Israel and Jordan.

~~~

Jordan could be said to be the tip of the profoundly intertwined triangle of conflict, colonialism and climate crisis. The issues that it faces have a significant bearing on the whole region, and how they are dealt with could come to dominate the area's political and environmental future.

Under Jordan and Palestine's faultline lies the separated African and Arabian tectonic plates. The Arabian Plate
~~~

houses what is called 'the most profound tectonic event of the earth crust', the Jordan Rift Valley. The Dead Sea, the lowest elevation point in the world, is here, as well as the River Aqaba[100] and, of course, the River Jordan, by whose banks the Prophet Muhammad is said to have been buried and in whose waters Jesus is said to have been baptised. It empties into the Dead Sea and shares its banks with Syria and Israel, due to the borders drawn up by the British in 1921. Much of the water is therefore siphoned off before reaching Jordan, which is a major reason why the country has one of the lowest levels of water availability per capita in the world, with residents of the capital, Amman, having access to water for just eight hours a week[101] and many rural communities living below the poverty line as agricultural work, reliant on water supplies, becomes scarcer. Yet Jordan's population grows exponentially yearly, as it provides relative safety for refugees from Palestine and Syria, meaning it has the second-highest share of refugees in the world compared to its population, the highest share being Lebanon's.

This water strain puts pressure on all resources, whilst the climate is changing quickly there, getting hotter and drier. There are those trying to establish plans for the Jordan Valley, which crosses the current borders of Israel, Jordan and Palestine, to become a 'vibrant green oasis'.[102]

EcoPeace is an NGO founded and run by Jordanian, Palestinian and Israeli environmentalists. Its slogan, 'Water Has No Borders', is geographically accurate, but not yet politically accepted, though it has been instrumental in pushing the urgency of environmental peace-building in the area, as one of the only organisations to attempt to put such theory into practice. Crudely, the theory argues that whilst environmental peace-building can't eliminate politics from decision-making, it can make technical factors more influential, building cooperation and trust as a result. There is arguably no better place than the Jordan Valley to do this. An energy exchange of electricity created by solar panels in Jordan for desalination of Israeli seawater is one of EcoPeace's projects that is currently being considered and points to cause for hope amongst otherwise rising tensions. EcoPeace's study concluded that, by 2050, Jordan supplying 20 per cent of the energy needs of Israel and Palestine would increase Jordan's GDP by 3–4 per cent, with total revenue flows allowing Jordan to purchase Mediterranean desalinated water at quantities enabling it to fully meet its own water needs and still be left with US$1 billion annually.[103]

For all its shifting tectonic plates, its political and geological rifts, historically this is an area of the water-stressed world in which hope holds out for a future of collaboration over conflict. It can be called hopeful, though not celebratory, due to the power inequality between its neighbours: Israel, which receives $3 billion per year from the USA purely for military spending; Palestine, which Israel illegally occupies; and war-devastated Syria. Jordan also receives $1.5 billion in US aid per year and is closely aligned with a number of European states for financial support to realise the Jordan Vision 2025, a strategy to improve Jordan's economy, ecology and society.[104] Any diplomacy that takes place under such conditions of inequity is unlikely to offer those with less bargaining power, such as Palestine and Syria, the very best solution for their problems. Despite this, in 2013, Israel, Jordan and the Palestinian Authority reached a historic tripartite agreement over the Red Sea–Dead Sea Conveyance Project.

The Red–Dead route proposes building a canal from the Red Sea at Aqaba through the Arava Valley to the Dead Sea, a distance of roughly 200 kilometres. After raising the water 200 metres to the peak of the Arava Valley eighty kilometres inland, the water would be allowed to run downhill through a series of cascades in order to generate electricity or hydrostatic pressure for desalination, or a dynamic combination of both. Supporters of the project also point out that the seawater could provide lakes and ponds along the route for recreation and aquaculture. The plans for desalination and the production of electricity would be used dynamically by season. During the winter, when demand for water is less than during the summer, desalinated water could be used to recharge depleted underground aquifers, or alternatively, to produce relatively greater amounts of electricity, saving on the use of oil and coal for power.

The project was designed to provide fresh water and electricity to the Jordan Valley via the desalination of Red Sea water, with the brine produced from the process being used to replenish the Dead Sea. Experts warn that at the current rate of depletion, the Dead Sea might have dried up by 2050, with devastating consequences for the region, especially Jordan. Environmental campaigners against the project warn that even if the project filled the Dead Sea, using desalination waste and 'mixing two seas' to do so could cause unprecedented ecological disaster. Transporting salt water across fresh-water aquifers would also be dangerous if there was leakage or a burst. The contamination of fresh water would be disastrous as underground aquifers are the only water sources for some regions and purification would be a difficult task. Building dams and artificial lakes along the route would also disrupt local ecosystems and cause displacement for local residents.

Nevertheless, the project has been researched, part funded and approved over the last seventeen years. Right now, relations have reached a historic low between Jordan and Israel, with Israel continuing to expand its illegal annexation of the Palestinian West Bank and Jordan Valley. The project has not yet begun due to these tensions, and Jordan recently announced its own new desalination plant, which has sparked rumours that hopes for the Red Sea–Dead Sea Project and its potential saving of the Dead Sea might be lost.[105]

~~~

Since the end of the Second World War and the creation of the State of Israel in 1948, Israel, Jordan and Syria have made competing claims for the River Jordan's water, carved up as it was by Sykes–Picot and then again by various Israeli-set boundaries which differed from those of Sykes–Picot and also from recommended UN Partition plans.[106]

When the colonial mandate for Palestine ended on 14 May 1948, the Jewish leadership declared the establishment of the State of Israel, without details of borders. The following day, armies from Egypt, Lebanon, Transjordan, Syria and Iraq began to fight with the Israeli Army until 1949, when armistice lines were drawn. These
~~~

became known as East Jerusalem and the West Bank, the latter occupied by Jordan, and the Gaza Strip, occupied by Egypt. The Arab states refused to recognise Israel, and so its borders remained unsettled until the 1967 Six Day War with Egypt, when Israel occupied the Egyptian Sinai, the Gaza Strip, the West Bank, East Jerusalem and the Syrian Golan Heights, tripling Israeli territory. The international community rejected the idea that these occupations had any legitimacy and refused to recognise Israel's border claims. In 1982, Egypt became the first state to make a treaty with Israel, regaining the Sinai and giving Israel its first official border line. Jordan followed in 1994, officially recognising Israel and formalising their shared border mainly through water, the boundary being the Jordan and Yarmouk rivers, the Dead Sea and the Gulf of Aqaba. In exchange, amongst other things, Israel was to give Jordan 50 million cubic metres of water per year and Jordan was to own 75 per cent of the water from the River Yarmouk.

~~~

Today, Israel's border with Syria remains unsettled and the West Bank and Gaza Strip are considered a single occupied entity by the UN and most of the world's governments.
~~~

Given this historical context and the current situation with Israel refusing to halt annexation, and Jordan being even more reliant on the annexation-supporting USA for aid ($1.3 billion in 2019) in a post-Covid world, despite all the efforts of groups such as EcoPeace, it seems highly unlikely that another tripartite agreement over water will be possible in the near future.[107] If the urgency of hydropolitics *can* manage to bring back functioning diplomatic relations, then it would be a way for states across the world to see water as a potential peace-builder, rather than a cause for conflict. Given the way more powerful states use, divert and steal water, this does seem idealistic, but as we hurtle into ever-increasing unknowns, we must hope and push for environmental and resource cooperation as a matter of emergency.

~~~

'Virtual water' is an area of collaboration that is becoming increasingly successful and popular. As it is based mainly on food trade, it has not yet been as affected by political relations between countries. Jordan is one of many already importing water virtually due to its water scarcity. This economically invisible and politically silent trade is accessible primarily through the international grain market. Given the current population of the Jordan basin,
~~~

the region would need about 15 billion cubic metres of water to be self-sufficient. There are less than 3 billion cubic metres of fresh water available annually. As grain crops would constitute the majority of that water usage, the Middle East is the world's biggest importer of grain, and this allows its water deficit to be managed.[108] Egypt is the biggest 'virtual water' importer in the area with 19.9 billion cubic metres in 2018, followed by Saudi Arabia, with 13 billion cubic metres. The Jordan Valley is third, with its 12 billion cubic metres of indigenous water deficit being provided by virtual water.[109] This is all possible due to mutual global trade. If the West was no longer dependent on oil, would the region survive the water scarcity?

Some think the answer lies not in regional cooperation or virtual water imports, but in water conservation and resource maximisation within states:

> The crisis relates fundamentally to the nature of water allocation and use within states rather than to water allocation between states. If hydro-diplomacy can start at home; by managing resources more carefully, the urgency of the water crisis will be a less-pressing issue of contention in Arab–Israeli relations.[110]

With conservation, the cost of saving water is almost always less expensive than the marginal cost of producing it elsewhere. As in the UAE, the current water shortage has forced Israel and Jordan into completely reassessing their use of water. There are numerous ways this can be achieved, such as agricultural water reallocation. This is when farmers take away water from farming and give it back to households. Switching to farming less water-needy crops has also been implemented across the region, growing the less water-dependent plants and then importing those foods that need a lot of water to grow. Drip irrigation – a drip system to water plants by means of plastic tubing with small holes in it – has also seen huge improvements in food production without increasing water usage.

~~~

Fresh-water management strategies, improved conservation techniques and slowed population growth have been put forward as the best plans for the Middle East's water issues. In Jordan, money has been allocated to giving instruction in women's health and to fledgling sex education lessons, primarily with the goal of saving water through slowed population growth. The current focus is on women's reproductive health choices and how to maintain a desired family size of their choice. This might
~~~

raise concerns about the integrity of wanting women to be empowered in their choices simply to serve the maintenance of a precious resource, but it is hopefully a step in the right direction for women to have more control over crucial aspects of their lives.

In domestic environments, scientists stress voluntary water conservation such as by making toilets, showers, and washing machines more efficient, by limiting outdoor use, and by repairing leaks, which can reduce domestic water consumption by as much as 62 per cent. Which is partly why, in a country with one of the lowest numbers of women in the workforce in the world, Jordan has been actively training a record amount of women to be plumbers. They hope to have the biggest plumbing force in the world by 2050. I imagine those superhero women with plungers, and I love them already . . .

***

LEAK! Leak! Is there a leak?
Can you see one?
Can you hear it,
the hiss of water wiggling
its way out into the world
uncaptured, wasted, wanton?

I won't let that happen.
Let the world know,
I stopped the leak.
Let the world know,
my fingers found the cracks
and their nimble little tips
fucking fixed them.
Superhero with a plunger.
Superhero with temporary plugging gum
making Jordan proud
to be the plumbing capital
of the Middle East, 2050.
Finally,
a capital of something,
not just leaks and refugees.
How do you like that,
those who said
I would only ever amount
to a pile of flesh
good enough to be married off
no later than nineteen –
now I am one of the blessed ones.
A plumber.
I have power now.

Power in ways it would have been crazy
to imagine before the Water Wars.
I flow through the roads unhindered,
my uniform like the fin of a Great White
moving ripples of fish away from my bite
nobody dares delay me,
distract me for just half a minute
and that might mean
no water for you tonight!
When the sun beams
I get sent ice creams
please come to me, to mine
don't let my pipes lose a drop!
I'm unstoppable.

There was a time when a woman
would never be able to bend
under a stranger's sink
and say, I can see where the problem is,
it'll take me ten minutes.
Pass me the spanner, please.
I mean, they physically could,
but the very phrase
'pass me the spanner, please'
would be interpreted sexually,

because every single thing we did
was interpreted sexually
because every single particle of power
was granted to us on the gravitas
of our sexual possibility
by those who may or may not
be sexually interested in us
but needed us to know that
we were only powerful
if they decided someone, somewhere
would be sexually interested in us eventually.
But now, they know we are the best,
that us women save more drops per hour
than any boy they'd ever honoured
with the Top Drop Hero Award,
maybe because our fingertips
are more sensitive
maybe because we had to work
harder for it
maybe because our eyes were trained
to search for drops
of whatever was being lost
from the day we were born –
whatever it was, that was it,
girls were picked

out of all the thousands of applicants
to be the senior plumbers,
the new emergency service
keeping society surviving –
wait, is that it? The leak?
No, just a dust rat,
damn tails sound like trickles
on these tiles.
Anyway, the Water Wars
were alright for some of us,
if you look at it like that.
Oh, look, I see it,
the leak the leak the leak
should have known, always look up,
a wobbly tear of water about to drop
from the ceiling –
CATCH!
Just like that.

***

Sitting in that interview room, staring up at the ceiling, listening to the drip drip drip of the tea and coffee machine, I knew my presence there was supposed to be a microscopic part of a hopeful, peace-building strategy of cross-cultural collaboration. I am still hopeful, or perhaps naive enough, to think that most of the people responsible for this strategy, for allowing me to join them up until the point of being in that interview room, were ignorant of how such a collaboration could never produce satisfactory results when the starting points were so inequitable and that inequity would be focused on as something *I* had to explain and justify to continue the collaboration. They never had to explain anything, and I accepted this should be the way of things, since that has always been the way of things: those in power never needing to explain or justify their power, but those without it having to scramble around in their every memory to think of reasons why they should be allowed to stoop even slightly adjacent to such power.

The interrogator told me the interview room was 'a safe space'. I didn't have the words then to articulate why I didn't think that it was, but I asked him something along the lines of how he could think it was a safe space for me, to be alone with a man in a room, considering the kinds of questions he'd asked, the answers I'd given. He didn't need to explain himself.

> *– Where do you go to feel safe, Sabrina? To feel supported and accepted, safe from external or internal danger. Do you go to the mosque?*

I used to visit the multi-faith prayer room at uni, get some spiritual nourishment alongside Hare Krishnas and Catholics and astrologists. Back then, most women's spaces in mosques weren't particularly inviting. In recent years I've felt more comfortable attending, if I ever do, with organisations such as Inclusive Mosque, who run prayers in pop-up locations to cater for all those looking to find acceptance and comfort outside of the traditional spaces.

– No, I don't.

– But you do pray, sometimes, to Allah?

– Yes.

– Do you ever seek advice from religious figures?

– If I said, 'Yeh, from a priest' would that be put down as a 'risk'?

– I can see you're getting agitated.

– I don't know any Islamic 'extremists' if that's what you're asking. I did tick 'no' to that question on the forms.

– I'm just attempting to build a picture of where you might find safety in a crisis situation. Would you meet friends for dinner, drinks?

– I'd rave, really, crisis or not, I go raving, sir.

– 'Raving'?

– Dancing. In big nightclubs, warehouses, jungle, garage, drum'n'bass –

– *You feel safety in raving?*

– Absolutely. Completely.

– *More so than in religion?*

– Yes, sir, what can I say, I was baptised in bass.

***

This is a safe space. What is a safe space? For whom is it safe? A safe space. Doesn't even always mean closed doors, soft chairs, sympathetic faces. For some of us, it will always mean big beats, meeting a stranger's eye seconds before a drop and feeling familial like my god we could have shared a womb during the same year. It will mean walls sweating ceilings dripping fingers gun-shaped and pointing, a reload making its way to the heart of the matter to our matter to what matters and what doesn't to the pain inside which is the pain we don't want to find ourselves faced with in a small space with a closed door, we might want magnitude, a huge cavernous roof which used to cover stored goods from an empire that stored us up to be released, here in this city, in this city that we would give ourselves over to entirely, entrusting it to be just that, to be just ours even though we knew better we knew when a mix was out of time we knew when the MC was too high to keep up, make sense, we knew we could never make sense here and yet, it was all we had so we leapt up and kept it as close as we could. Our safe space. Raving, always. Wherever we are.

***

For me, it was mostly in London. But I'd raved in Ramallah, all over Egypt, Istanbul, Dubai and Jordan. Surprisingly, it was in Jordan that I first properly raved to Egyptian electronic music. The raves in Egypt that I'd been to had heavily leant to American and European house music. I was in Jordan as part of the World Economic Forum's MENA (Middle East and North Africa) Conference, as a 'Global Shaper', a group of under-thirty-year-olds who brought some wide-eyed youth to a conference with an average participant age of seventy-five, and in return got a nice hotel room and access to the world's senior politicians and business people. But, as is the case with wide-eyed youth, despite this privileged proximity to power, we were mostly looking for the party. When we found it, the sounds were raw synthetic beats embroidered with syncopated tabla samples and queasy electronic loops. The voices of the young men in skin-tight jeans jumping around the mixing board auto-tuned to a sharp metallic edge. I was told this was a new genre called Mahragan, now sometimes described as an Egyptian version of rap or grime, but it is the digital-age descendant of the 1970s Egyptian *sha'abi* songs of working-class woes, of loves and jobs lost. Bawdy and festive, sha'abi helped boost the national mood in the wake of Egypt's 1967 defeat by Israel. It was the dawn of the cassette era in Egypt, and sha'abi spread freely on bootlegged tapes. The internet has allowed Mahragan to go much further, as both a means of production thanks to free downloadable software, and of promotion and distribution via YouTube, SoundCloud and Facebook. It was officially banned by the Egyptian authorities in 2020, due to the 'promiscuous and immoral lyrics'.[111] But at the time when I was in Jordan, its novelty as an Arab sound was surging uncontrollably through the underground of the region, each country adding its own twist to it. Maybe it was then, or maybe I'd always known it, but as the crowd danced happily to collective transcendence, I realised the only cross-cultural collaboration I'd ever be accepted within, and the only one I would want to be a part of, would be in the worlds of music, writing, theatre, art. Is it in these worlds that the real difference can be made? I've still not come to a conclusion on that, but I was right about the acceptance part. Whilst Jordan offers international politics a real-world example of environmental peace-building being attempted, that evening it offered me the possibility of rebuilding the ways we are taught to be 'safe', through the means of artistic expression and solidarity. The ideas to which I eventually committed, as a way of living and working and pushing for change. Ensuring that I could put what I ask of myself and society before what a job asks of me. If I'd been able to fit easily inside the hard drawn lines of the Top Secret Clearance borders, this might never have happened. So for that, in a way, I am grateful.

SCOTLAND
NORTH SEA
IRISH SEA
ENGLAND
WALES
London
ENGLISH CHANNEL
FRANCE

Harbour II

(England)

I have here that you attended an Arabic-language school in Cairo.
Why would you do that?
Didn't you learn it at home?
Why would you go to Arabic school when you have an Arabic dad?
Answer me that.

~~~

Home. Sand. Hand. Flat. Break. Build. Break. Build. Break.

Build. Break again. Build again. When will it end? Will it end?

Sand in your face. Grains in your eyes. I want you to go blind. Sting the arrogance from your sight, if that is even possible. Want to watch you struggle. Want to watch you flail to get the tiniest broken-down bits of stone out from the roving globes inside your head, see how hard it is, to see when you are unable to see, when something unbeatable has come along to block what you
~~~

have been given to see with, weep, weep, weep all the water you can, they will not be washed out, they will stay and you will have to find a new way to be. That is what I want to do. Take handfuls of this sand and grind it into your sockets. Feel the rough particles push themselves into plasma, imagine how you'd shout and scream trying to maintain decorum, some kind of superior feeling of having been torture-trained and trained in torture but it won't matter because I won't stop and eventually you will kneel and beg and beg and beg and I will push your forehead up and say look at me and –

– He didn't want me to speak Arabic, sir. I used to be angry about that, but now I think, he just didn't want you to do *this*.

– ***And why do you, Ms Mahfouz, want to do this?***

Can you believe in a place? A piece of land? An expanse of water contracted to a country by another country? I believe in the Middle East. I believe in MENA. Middle East and North Africa. Or the more geographically accurate SWANA, South-West Asia and North Africa. I believe in the way people say they do to get to write books like this one, but actually what does that even mean? I believe, as in I believe it exists as an actual land mass? And, if so, is this because the map says it does or because I've taken trips, I've lived within its limits, I know that it is really truly part of the world and so I believe in it, it is not a myth, it is not just grains of sand in an hourglass blown by the great European men who showed it to us? Or as in, I believe in its potential to be better, its potential to teach, its potential to reach its potential? Or as in, I believe in its heart, how it beats inextricably with my own? I believe in London too. In all these ways. Even affording it such specificity as its own name for a 600-odd square mile radius, whilst giving MENA an acronym to cover a quarter of the world's land mass, shows how I have held my city, how the world has been made to hold it, whether they believed in it or not.

~~~

Whilst the British were abroad building their empire in every corner of the world, what was Victorian London like? It was a sprawling metropolis, the nerve centre of the most imperial nation on earth at the time. Yet it was also a city wracked by fear. Confronted by rapid industrialisation, crowded slums and graphic media reports of a thriving criminal underbelly, many Londoners may have come to the conclusion that danger lurked in every shadow. Including in the shadows of the sewers. The
~~~

city's nascent sewer system was overtaxed and constantly breached its limits. The network was overhauled and markedly expanded after the Great Stink of 1858, when deposits that the sewers had spat out into the River Thames baked in the summer heat, to the stomach-roiling chagrin of the entire city.

The journalist Henry Mayhew was unfortunately familiar with the contents of the city's subterranean guts. He spent the 1840s documenting the lives of the city's mudlarkers, ratcatchers, food vendors and other working folks, and the sewers came up a lot in his series, which was compiled into a multi-volume set, *London Labour and the London Poor*, first published in 1851. During the spring tides, Mayhew wrote, fetid liquid 'burst up through the gratings into the streets' until the low-lying neighbourhoods around the Thames 'resembled a Dutch town, intersected by a series of muddy canals'. He also spoke to sewer hunters, or people who made their livings on other people's leavings. To hear Mayhew tell it, they were busy. 'Some few years ago, any person desirous of exploring the dark and uninviting recesses,' Mayhew wrote, could walk right in one of the pipes that emptied into the Thames 'and wander away, provided he could withstand the combination of villainous stenches which met him at every step, for many miles, in any direction.'

They travelled in groups to defend themselves from rats, strapped lanterns to their chests, and raked the mud with long-handled hoes to find any money, nails or metal scraps that may have got lost and lodged down there. They told stories they swore were true about packs of killer sewer pigs that rampaged underneath the streets of London. How a pregnant sow had roamed into the sewer somewhere near Hampstead, and then delivered and raised her offspring in the pipes. Feasting on 'the offal and garbage washed into it continuously, the breed multiplied exceedingly, and have become almost as ferocious as they are numerous'.

~~~

I visited a disused London sewer as research for a filming location a few years ago. We passed by a small fold-up stool and the guide said whilst there are no official figures, it is assumed there are people who live in the sewer system due to the paraphernalia of everyday living that is often found.

I couldn't stop thinking about the sewers of London being renewed and extended with the money being made from the exploitation of the Middle East's waters, the unexpected legacies of that in a country vulnerable to
~~~

flooding. And yet the UK is liable to run out of water in twenty years if the privatised water companies don't fix the leaks that they've neglected over the last two decades, which have resulted in 3 billion litres of water being lost per day. This preventable wastage is double the amount of daily water that Jordan uses overall.[112]

Most of England's rivers never satisfied EU ecological standards, with 0 per cent of them meeting a 'good chemical or ecological status', and due to Brexit the government will no longer be legally obliged to address it.[113] The UK is one of the most nature-depleted countries in the world, with over half of its native species in decline, which is greatly related to water issues.[114]

There is so much to be angry about, from both the past and the present, both politically and environmentally – inextricable as they are, one from the other. But I know that like many others who are angry, I can still love this country, even though it is built on the blood of empire and slavery and the ashes of slum bodies burnt in fires and plague pits and priceless bits of other people's lives. I can hope and push and challenge it to take its present into a more hopeful future by accounting for its past.

~~~
~~~

I think of its future and I think of its water.

Will mismanagement mean it is parched? Or will it be flooded, will it have too much water, will that be better than those who will not have enough, in part because of its history? Or will we one day have a government that doesn't fear honest, deep reflection on Britain's past and how it perpetuates colonial policies today, the dire long-term consequences of that? That looks at the water available to the UK, which ranks 102 out of 179 countries for fresh-water resources, far below a number of African and South American countries,[115] and decides to prioritise its management over private profits? Perhaps plans to manage its water so well it can offer its excess to those much further down the list, the final ten all being countries in the Middle East that Britain was once directly administering and exploiting?

I think of the UK's opposing water threats of the present and future, drought versus floods. It seems fitting somehow. The way it gives and takes away, never getting the balance right. I think of those who may arrive, fleeing drought then hoping not to be drowned. Of those who are only ever given the choices afforded to them by empire and its legacies.

Whilst rehearsing my show that became the inspiration for this book, *A History of Water in the Middle East*, at the Royal Court Theatre, my friend Kareem Samara, who was also the composer and musician in the show, pointed out the irony of what we were doing, considering the very space we were occupying at that moment. The rest of the group looked at him quizzically and expectantly; it had become a common occurrence that Kareem would look up from his pedal loop and his oud-playing to deliver facts about our shared city that had a tendency to re-pivot all we had thought we knew about what we were making and where we were living. This time was no exception. 'Hans Sloane, innit?' Now he looked at us quizzically, expectantly. We had no idea who he was talking about.

The Royal Court Theatre presides over a corner of Sloane Square, prime real estate in one of the most expensive areas of London and the world, housing what we theatre-makers liked to regard as a radical and renegade laboratory for delivering shows that sell tickets but also pushed boundaries and buttons, especially if it was to go on in the smaller attic space above the grand main theatre, as ours was programmed to do. Kareem told us that Hans Sloane, whom Sloane Square was named after, was the founder of the British Museum. As a colonial officer, he 'brought back' thousands of artefacts which formed the core inaugural collection of the museum. He also commissioned much of the building of the SW1 area of London, including the square where we, all products of empire, now sat rehearsing a show singing its crimes. Sloane's wife was the one with the money which enabled him to do this building and 'collecting'. Her family were immensely rich from Jamaican sugar plantations, from slavery. The direct descendants of that family are the Cadogans, who now collectively own ninety-three acres of land in Kensington and Chelsea, Earl Cadogan and his family having a net worth of £6.85 billion in 2019.[116]

Oh. We sat silent. Any politically engaged and historically informed person who has grown up in London is used to most central streets, buildings and landmarks having imperial links, existing only because the empire did. For the majority of us in that rehearsal room, we only existed because the empire did. Our parents were immigrants from ex-British colonies, meeting either in London or 'back home' and travelling to London to make a new life because they had grown up thinking of it as their capital, no matter how many thousands of miles away it was, or because they had to. Our bodies were as much a marker of the British Empire as the grey statue of Hans Sloane that adorned the tree-lined square we ate our lunch in. But we owned nothing. We would not be made into marble memorials for our service. We would not be asked to update the coloniser's treasure haul by emptying our shelves of the mementos that mark British immigrant existence – miniature flags, gilded ashtrays, plastic-covered seats, rugs on top of carpets, photo frame after photo frame. But we had given parts of ourselves to build this place. Ancestrally and presently. And as artists, we *were* emptying out those mementos, with our music and words and dance that referenced identities we were not even sure we could claim any more, because when had anybody last visited 'home'? Who could speak the language as a mother tongue? How would we know our way around even one corner of the now-sprawling cities of our parents without satnav? It had felt like this theatre piece was being made on our terms. We were there because we had worked hard to be, and we just happened to be telling the stories we wanted to tell. The comment from Kareem changed that. Nothing we did in this country could ever be on our own terms. The terms of our engagement with it and its engagement with us were set centuries ago. Yes, in some ways they have shifted with the nuance of mixes and increased understanding of the legacy of empire, but in some ways, they have not, no matter how many statues fall.

~~~
~~~

– *Why do* you, *Ms Mahfouz, want to do* this?

~~~

Maybe I have been conditioned to see myself anthropologically, just like you. Orientalism internalised, through and through. Maybe I am desperate, so very very desperate, to prove to myself that I do contain multitudes, that I am not just a vessel of cultural interest with no interior life, that I want it all. I want to be so stuffed with supposed contradictory experiences that I can never be studied satisfactorily. I can never be commented upon without those Orientalist clichés you didn't even know you had rising up to condemn my brazen ways of wanting to be too many things, in too short a time, 'lifestyle choices' you might call them. 'Attempts at internal decolonisation' would be my preferred phrase. But I need the fantasy, don't I? To be able to grab these things and stuff them inside me, to fill myself up to bursting beyond a stereotype, I need your fantasy to do it and maybe I will be completely lost without it. Being right here, with you, inside the very centre of a reverberating force that has long outlasted its batteries, maybe I thought I'd finally win a way to see myself without you seeing me. Maybe I could only do that by replacing you, all of you. Maybe
~~~

that is why I wanted to do this. The interview room was silent, the tables stained with tea.

~~~

The next time we meet is in the lobby of the Ministry of Defence. His briefcase remains closed as he leans in and lets me know that, although to him I seem like a girl who genuinely just wants to do what good she can in the world, he had to assess on the facts that he had and the facts that he had made my risk factor extremely high.

> *– You understand, don't you, Sabrina, with your heritage and the corresponding warmth you feel for the Middle East and, of course, your debts, your lack of assets, financially, you understand, your lifestyle choices, well, bribery, it's a big risk with what we do.*

And there was no harm in trying again when things were different, he said, as he left through the capsule doors we keyed in and out of every day like in an episode of *Spooks*.

The official letter followed a few weeks later, less candid than the man was.

> Developed Vetting has not been granted and therefore you will not be able to progress into any roles which
~~~

require a higher level of security clearance than that which you already have, thank you.

~~~

All of that to get to this. This. Literally this.

You and me and this page. A page that fits me and you and all of this within it but it is still only one page. It is still only here. Inside. I can be big here, I can feel my words land across the whiteness, landing in your eyes, and I imagine how you might make them your own when you tell people what you read when you came here, how my words will become your words and really that is such a huge power, to be able to literally put words into people's mouths, maybe that is what writing and speaking are, just regurgitations into your own intonations of everything you've ever read or seen or heard and if so then words are like water, we drink it bathe in it eat it consume it piss it back out and it comes back to us, never lets us go, it contains everything we have ever done or said or been and so when somebody like me leaves a page like this, we can drown. We can drown because there is too much that needs to be said in our own words, but we have to drink the other ones first before we can say our own and lungs can't cope with that much liquid it's just not how it works your lungs need space you need
~~~

space in there to make the new air and air and water just make bubbles and bubbles keep the air caged and then the water will always win, it always wins, you cannot fight the tide. So when I go outside and I am no longer here in front of you all I can do is drink. And now I have, if I'm lucky, half a life left I must set out my drinking habits sensibly. I must set out my preferences, my bottled-up boundaries. I need as many others as possible to be making words their own, so the drinking is replenishing, hydrating, does not choke. Just as we have to change the way water is consumed, managed and recycled, so we have to change the way we drink in words.

~~~

I never got to say –
you made me suspicious of myself, who I could become without being a part of you, assessed and monitored by you. I wanted to save myself and save you by giving myself to you. But you didn't want me –
So I say it now.

~~~

Of course, there would be no reapplication. No reassessing of how much risk the regurgitated water of my life poses to those who have never had to spit. There will be no more sitting, listening to myself told to me by someone who is not me telling me that I am risky, that by being a product of untamed waters I will never know internal order; I will never have the wherewithal to know Whitehall's wrong from right to truly align the fight inside me.

~~~

What nobody ever said directly was the truth of course.

We know you will uncover an anger underneath the waterbed when you realise how we regard you, which you will when you grow and rocks are thrown, you will see, and when you do, we know what you can do, we have decimated and destroyed and maimed and raped and ripped apart to make sure you can't so of course we can never trust you, because one day, you will see.

And they would have been right. I did.
~~~

Acknowledgements

I am so grateful to everyone who supported my initial creative journeys, both literal and metaphorical, into the waterways of the Middle East, primarily Orla O'Loughlin, Vicky Featherstone, Arts Council England and Shubbak Festival. This support allowed me to begin delving deeper into the imperial history of these waters and eventually to write the play, *A History of Water in the Middle East*, which was the start of this book. That play wouldn't have happened without the team who made it, so massive thanks and love to all of them – Stef O'Driscoll, Kareem Samara, Laura Hanna, David Mumeni, Masha Kevinova, Maria Koripas, Khadija Raza, Prema Mehta, Dominic Kennedy, Charli Davis, Lavinia Serban, Sophia Dalton, Bex Kemp and everyone at the Royal Court Theatre.

This book would not have been a book without the stirring support and belief of my editor Amy Perkins, all the team at Tinder Press and my agent Becky Thomas, thank you! And a special mention for the initial push from Ellie Carr.

As always, thanks to my friends and family, especially Tabari, who puts it all in perspective. Extra thanks to my mum for her help with researching personal and historical aspects of our itinerant life together, and to Anthony for emboldening me to write what and how I need to.

Gargantuan gratitude to all the writers I have had the privilege to read, work with and get to know, for all the inspiration, consolation and encouragement they've shared with me. There's too many to list, but you're on my shelves and in my heart. Special thanks to all the Egyptian writers who continue to fire me up, calm me down and educate me along the way – Omar Robert Hamilton, Ahdaf Soueif, Yasmine El-Rifae, Alaa Abd Al-Fattah, Salma El Wardany, Alya Mooro, Ramy Youssef, Mona Eltahawy and my great grandfather, Ahmed Mahfouz.

Unending shout outs to all the courageous, scared witless, silent, loud women who fight in their own ways for their stories to be heard and to be changed, or accepted.

Mostly, and finally, I want to acknowledge all the writers who put their freedom and lives at risk to document all aspects of the world we inhabit. Not just the explicitly political, but the personal, the sexual, the spiritual. Without them, words would no longer hold truth.

Further Reading

A Decolonial Feminism; Françoise Vergès, 2021, Pluto Books

A Line in the Sand: Britain, France and the Struggle That Shaped the Middle East; James Barr, 2011, Simon & Schuster UK

Assuming Boycott; Ed. Kareem Estefan, Carin Kuoni, Laura Raicovich, 2017, OR Books

Border and Rule: Global Migration, Capitalism and the Rise of Racist Nationalism; Harsha Walia, 2021, Haymarket Books

Border Nation: A Story of Migration; Leah Cowen, 2021, Pluto Press

(B)ordering Britain: Law, Race and Empire; Nadine El-Enany, 2020, Manchester University Press

Classical Poems by Arab Women: A Bilingual Anthology; Ed. Abdullah Al-Udhari, 1999, Saqi Books

Coming of Age in the War on Terror; Randa Abdel Fattah, 2021, NewSouth Publishing

Cut From the Same Cloth: Muslim Women on Life in Britain; Ed Sabeena Akhtar; 2021, Unbound

Empire's Endgame: Racism and the British State; Ed. Gargi Bhattacharyya, Adam Elliot-Cooper, Sita Balani, Kerem

Nisancioglu, Kojo Koram, Dalia Gebrial, Nadine El-Enany, Luke Noronha, 2021, Pluto Press

Feminism, Interrupted: Disrupting Power; Lola Olufemi, 2020, Pluto Press

Hidden Heritage: Rediscovering Britain's Lost Love of the Orient; Fatima Manji, 2021, Chatto & Windus

Midnight in Cairo: The Female Stars of Egypt's Roaring 20s; Raphael Cormack, 2021, Saqi Books

Migrant City: A New History of London; Panikos Panayi, 2020, Yale University Press

Mixed/Other: Explorations of Multiraciality in Modern Britain; Natalie Morris, 2021, Trapeze

Natives: Race & Class in the Ruins of Empire; Akala, 2018, Two Roads

Revolutionary Feminisms; Ed. Brenna Bhandar and Rafeef Ziadah, 2020, Verso

Tangled in Terror: Uprooting Islamophobia; Suhaiymah Manzoor-Khan, 2022, Pluto Press

The Brutish Museums: The Benin Bronzes, Colonial Violence and Cultural Restitution; Dan Hicks, 2021, Pluto Books

The Crusades Through Arab Eyes; Amin Malouf, 1984, Saqi Books

The City Always Wins; Omar Robert Hamilton, 2017, Faber

The Greater Freedom: Life as a Middle Eastern Woman Outside the Stereotypes; Alya Mooro, 2019, Little A

The New Age of Empire: How Racism and Colonialism Still Rule the World; Kehinde Andrews, 2021, Penguin

The Nile: Travelling Downriver Through Egypt's Past and Present; Toby Wilkinson, 2014, Vintage

The Responsible Globalist: What Citizens of the World Can Learn From Nationalism; Hassan Damluji, 2019, Allen Lane

We Have Always Been Here: A Queer Muslim Memoir; Samra Habib, 2019, Riverrun

We Wrote in Symbols: Love and Lust by Arab Women Writers; Ed. Selma Dabbagh, 2021, Saqi Books

When We Were Arabs: A Jewish Family's Forgotten History; Massoud Hayoun, 2019, The New Press

White Tears/Brown Scars: How White Feminism Betrays Women of Colour; Ruby Hamad, 2020, Trapeze

You Have Not Yet Been Defeated; Alaa Abd Al-Fattah, 2021, Fitzcarraldo Editions

Bibliography

1 https://assets.publishing.service.gov.uk/government/uploads/system/uploads/attachment_data/file/767789/Civil_Service_Fast_Stream_Annual_Report_2017_-_2018.pdf.

2 https://www.historic-uk.com/HistoryUK/HistoryofBritain/Timeline-Of-The-British-Empire/.

3 R. Cox, 2017 , 'Expanding the History of the Just War: The Ethics of War in Ancient Egypt', *International Studies Quarterly*, vol. 61, no. 2, June 2017, pp. 371–84. https://doi.org/10.1093/isq/sqx009.

4 https://www.historic-uk.com/HistoryUK/HistoryofBritain/Timeline-Of-The-British-Empire/,

5 E. Helal, 'Muhammad Ali's First Army: The Experiment in Building an Entirely Slave Army', in T. Walz and K. Cuno (eds.), *Race and Slavery in the Middle East: Histories of Trans-Saharan Africans in Nineteenth-Century Egypt, Sudan, and the Ottoman Mediterranean* (The American University in Cairo Press, 2010), pp. 17–42.

6 https://www.theguardian.com/global-development/2020/nov/12/dozens-of-sudanese-migrants-held-in-cairo-after-protests.

7 https://www.oxfam.org/en/press-releases/more-52-million-people-across-africa-going-hungry-weather-extremes-hit-continent.

8 https://asia.nikkei.com/Spotlight/Asia-Insight/Flooded-Asia-Climate-change-hits-region-the-hardest.

9 https://www.bloomberg.com/graphics/2019-countries-facing-water-crisis/.

10 Marty Gould, *Nineteenth Century Theatre and the Imperial Encounter* (Routledge, 2011).

11 Angus Konstam, *Nile River Gunboats 1882–1918* (Osprey, 2016), p. 4.

12 https://www.forconstructionpros.com/blogs/construction-toolbox/blog/12096401/looking-back-on-the-worlds-deadliest-construction-projects.

13 http://webs.bcp.org/sites/vcleary/modernworldhistorytextbook/imperialism/section_6/suezcanal.html.

14 https://ukdefencejournal.org.uk/british-troops-deploy-to-egypt/.

15 Benjamin Poore, *Theatre and Empire* (Palgrave Macmillan, 2016).

16 https://www.iwm.org.uk/history/why-was-the-suez-crisis-so-important.

17 O. Smolansky, 'Moscow and the Suez Crisis, 1956: A Reappraisal', *Political Science Quarterly*, 80(4), 1965, pp. 581–605. doi:10.2307/2147000.

18 https://www.iwm.org.uk/history/why-was-the-suez-crisis-so-important.

19 https://www.gov.uk/government/publications/exporting-to-egypt/doing-business-in-egypt-egypt-trade-and-export-guide.

20 https://see.news/uk-tradeenvoy-to-egypt-discusses-investment-in-suez-canal/.

21 https://yemen.unfpa.org/en/news/three-million-women-and-girls-risk-violence-yemen.

22 https://www.ohchr.org/EN/HRBodies/HRC/Pages/NewsDetail.aspx?NewsID=24937&LangID=E.

23 https://www.theguardian.com/world/2018/aug/19/us-supplied-bomb-that-killed-40-children-school-bus-yemen.

24 https://www.independent.co.uk/news/uk/home-news/uk-saudi-arabia-arms-sales-yemen-war-weapons-treaty-oxfam-a9258166.html.

25 https://www.theguardian.com/law/2019/jun/20/uk-arms-sales-to-saudi-arabia-for-use-in-yemen-declared-unlawful.

26 https://www.independent.co.uk/news/uk/home-news/yemen-civil-war-british-weapons-poll-human-appeal-half-unaware-saudi-air-strikes-civilian-deaths-a7636271.html.

27 *Encyclopedia of Women & Islamic Cultures: Family, Law & Politics*, ed. Suad Joseph (Brill, 2005).

28 https://www.marxist.com/colonial-revolution-and-civil-war-in-south-yemen.htm.

29 https://muslimhands.org.uk/latest/2021/04/providing-water-to-almost-2-million-people-in-aden-yemen.

30 https://nationalinterest.org/blog/middle-east-watch/beyond-suez-how-yemen's-war-can-imperil-red-sea-shipping-181647.

31 https://commonslibrary.parliament.uk/yemen-uk-governments-aid-reduction/.

32 https://hansard.parliament.uk/commons/2021-03-02/debates/62AC3E67-7B7A-4B94-898A-8D18B55A7A1C/YemenAidFunding.

33 John Gunther, *Inside Europe* (Hamish Hamilton, 1940).

34 https://www.theneweuropean.co.uk/brexit-news/nigel-farage-on-the-civil-service-and-military-49688.

35 https://publications.parliament.uk/pa/cm201314/cmselect/cmfaff/88/8808.htm.

36 https://hansard.parliament.uk/Commons/2018-11-20/debates/F280FC05-9A95-4F2E-BAEC-E2AE46190703/RoyalNavyBaseBahrain.

37 https://hansard.parliament.uk/Commons/2018-11-20/debates/F280FC05-9A95-4F2E-BAEC-E2AE46190703/RoyalNavyBaseBahrain.

38 https://interactive.aljazeera.com/aje/2016/sykes-picot-100-years-middle-east-map/index.html.

39 F. Ahmad, 'A Note on the International Status of Kuwait before November 1914', *International Journal of Middle East Studies,* 24(1), 1992, pp. 181–5.

40 R. Pillai and M. Kumar, 'The Political and Legal Status of Kuwait', *The International and Comparative Law Quarterly,* 11(1), 1962, pp. 108–30.

41 Ahmad, 'Note on the International Status of Kuwait', pp. 181–5.

42 Dalal Al-Harbi, *Prominent Women in Central Arabia* (Ithaca Press, 2008).

43 https://www.researchgate.net/publication/300192057_Fashioning_the_Future_The_Women's_Movement_in_Kuwait.

44 https://www.pionline.com/economy/oil-rich-kuwait-running-out-cash.

45 Maryann Tétreault, Helen Rizzo and Doron Shultziner, 'Fashioning the Future: The Women's Movement in Kuwait', in Pernille Arenfeldt and Nawar Al-Hassan

Golley, eds., *Mapping Arab Women's Movements: A Century of Transformations from Within* (The American University in Cairo Press, 2012), pp. 253–78. doi 10.5743/cairo/9789774164989.003.0010.

46 https://vimeo.com/125652842.

47 https://www.statista.com/statistics/478299/united-kingdom-uk-exports-value-trade-goods-to-kuwait/.

48 https://www.bbc.co.uk/news/uk-54435335.

49 https://www.bbc.co.uk/news/uk-politics-53324251.

50 https://ukdefencejournal.org.uk/the-uk-in-the-persian-gulf-historical-involvement-and-military-presence/.

51 https://www.privacyshield.gov/article?id=Kuwait-Oil-and-Gas.

52 https://www.britannica.com/event/Persian-Gulf-War.

53 https://www.britishlegion.org.uk/stories/the-gulf-war.

54 https://archive.globalpolicy.org/previous-issues-and-debate-on-iraq/41759.html.

55 https://sites.tufts.edu/reinventingpeace/2013/06/13/iraqs-assault-against-the-kurds/.

56 https://www.armscontrol.org/act/2006-01/report-confirms-iraq-used-sarin-1991.

57 Frantz Fanon, *The Wretched of the Earth* (Mcgibbon & Kee, 1965 [*Les Damnés de la terre*, François Maspéro, 1961]).

58 https://www.thebookseller.com/news/mackesys-boy-mole-fox-and-horse-tops-2020-chart-1232513.

59 https://www.amazon.co.uk/gp/bestsellers/2020/books.

60 https://www.militarytimes.com/news/your-military/2020/12/28/new-in-2021-us-troop-presence-heading-down-to-2500-in-iraq-and-afghanistan/.

61 https://www.opendemocracy.net/en/opendemocracyuk/britain-is-world-centre-for-private-military-contractors/.
62 https://www.militarytimes.com/news/your-military/2020/12/28/new-in-2021-us-troop-presence-heading-down-to-2500-in-iraq-and-afghanistan/.
https://www.stripes.com/news/middle-east/airstrikes-pummel-isis-in-iraq-as-us-prepares-for-talks-with-baghdad-1.667199.
63 http://sedici.unlp.edu.ar/bitstream/handle/10915/41162/Documento_completo.pdf?sequence=1.
64 https://www.un.org/press/en/2001/unep98.doc.htm.
65 https://www.bp.com/en/global/corporate/who-we-are/our-history/early-history.html.
66 http://www.danielpipes.org/164/a-border-adrift-origins-of-the-iraq-iran-war.
67 https://climate-diplomacy.org/case-studies/iraq-iran-water-dispute-war.
68 http://waterinventory.org/sites/waterinventory.org/files/chapters/Chapter-05-Shatt-al-Arab-Karkheh-and-Karun-Rivers-web_0.pdf.
69 https://www.nature.com/articles/s41598-020-63893-w.
70 http://www.ejolt.org/2013/08/declaration-of-the-ekopotamya-network-turkey-iran-iraq/.
71 https://www.vice.com/en/article/exmejj/iraqi-scouse-brow-292.
72 https://www.middleeasteye.net/features/iraqs-traumatised-women-seek-botox-alternative-therapy.
73 http://jdeedmagazine.com/calling-iraqi-diaspora-artists-the-idcn-is-a-new-digital-home/.
74 From *Layla and Majnun* by Nizami Ganjavi, translated by Colin Turner, John Blake, 1992.

75 https://www.btselem.org/freedom_of_movement/checkpoints_and_forbidden_roads.

76 Yonatan Mendel, *The Creation of Israeli Arabic: Security and Politics in Arabic Studies in Israel* (Palgrave Macmillan, 2014), p. 188: 'The exact percentage of Jews in Palestine prior to the rise of Zionism is unknown. However, it probably ranged from 2 to 5 per cent. According to Ottoman records, a total population of 462,465 resided in 1878 in what is today Israel/Palestine. Of this number, 403,795 (87 per cent) were Muslim, 43,659 (10 per cent) were Christian and 15,011 (3 per cent) were Jewish (quoted in Alan Dowty, *Israel/Palestine* (Cambridge: Polity, 2008), p. 13). See also Mark Tessler, *A History of the Israeli–Palestinian Conflict* (Bloomington, IN: Indiana University Press, 1994), pp. 43 and 124.

77 https://www.hrw.org/world-report/2021/country-chapters/israel/palestine.

78 https://www.un.org/press/en/2016/sc12657.doc.htm.

79 https://reliefweb.int/report/occupied-palestinian-territory/humanitarian-action-children-2021-state-palestine.

80 https://whc.unesco.org/en/tentativelists/5722/.

81 https://www.mashallahnews.com/the-excrement-river-of-gaza/.

82 https://www.ohchr.org/EN/HRBodies/HRC/RegularSessions/Session48/Pages/48RegularSession.aspx.

83 https://reliefweb.int/report/occupied-palestinian-territory/focus-effects-israel-s-tightened-blockade-economic-and.

84 https://www.haaretz.com/israel-news/.premium-israel-holds-up-vital-spare-parts-for-gaza-s-water-and-sewage-systems-1.10523771.

85 https://www.al-monitor.com/originals/2021/05/israeli-campaign-gaza-worsens-water-crisis.

86 https://www.amnesty.org.uk/gaza-operation-cast-lead.

87 https://www.aljazeera.com/news/2021/10/12/gaza-undrinkable-water-slowly-poisoning-people.

88 https://www.reuters.com/world/middle-east/sewage-dump-nature-reserve-un-hopes-save-gaza-valley-2022-02-07/.

89 https://reliefweb.int/report/occupied-palestinian-territory/20000-palestinians-benefit-clean-water-tanks-gaza.

90 https://www.irishtimes.com/news/ireland/irish-news/how-israel-used-desalination-to-address-its-water-shortage-1.3959532.

91 https://www.un.org/unispal/document/auto-insert-195880/.

92 https://www.jadaliyya.com/Details/41686.

93 https://www.forbes.com/sites/dominicdudley/2017/05/08/middle-east-worlds-deadliest-region/?sh=51bfbffa1744.

94 https://themarkaz.org/magazine/drought-and-the-war-in-syria.

95 https://worldpopulationreview.com/country-rankings/energy-consumption-by-country.

96 https://www.renewableenergymagazine.com/emily-folk/is-there-a-link-between-renewable-energy-20190924.

97 https://gulfnews.com/uae/environment/trial-run-for-uae-iceberg-project-in-2019-1.2244996.

98 https://www.icj-cij.org/public/files/case-related/131/1559.pdf.

99 Avi Shlaim, *Lion of Jordan; The Life of King Hussein in War and Peace* (Allen Lane, 2007).

100 Kamal Jreisat and Tawfiq Yazjeen, 'A Seismic Junction', in Myriam Ababsa, ed., *Atlas of Jordan: History, Territories and Society* (Presses de l'Ifpo, 2013), pp. 47–59.

101 Kayla Ritter, 'Amman Faces Water Squeeze as Refugees Rush into Jordan', *Waternews*, Circle of Blue, 3 May 2018, https://www.circleofblue.org/2018/middle-east/amman-faces-water-squeeze-as-refugees-rush-into-jordan/.

102 https://www.news957.com/world/2017/10/31/jordan-water-crisis-worsens-as-mideast-tensions-slow-action/.

103 https://old.ecopeaceme.org/wp-content/uploads/2020/12/A-Green-Blue-Deal-for-the-Middle-East.pdf.

104 https://www.greengrowthknowledge.org/national-documents/jordan-2025-part-ii.

105 https://www.al-monitor.com/pulse/originals/2020/12/jordan-israel-dead-sea-project-environment-threats.html.

106 A. A. Amery and A. T. Wolf, eds., *Water in the Middle East: A Geography of Peace* (University of Texas Press, 2000).

107 https://www.brookings.edu/blog/order-from-chaos/2020/06/01/with-israels-annexation-plans-looming-an-hour-of-decision-for-jordan/.

108 Marc Sadler and Nicholas Magnan, 'Grain Import Dependency in the MENA Region: Risk Management Options', *Food Security*, 3, 2011, pp. 77–89. doi: 10.1007/s12571-010-0095-y.

109 https://hess.copernicus.org/preprints/hess-2018-4/hess-2018-4.pdf.
110 Committee on Sustainable Water Supplies for the Middle East, *Water for the Future: The West Bank and Gaza Strip, Israel, and Jordan* (National Academies Press, 1999), pp. 100–116.
111 https://www.middleeasteye.net/news/egypt-musical-syndicate-bans-popular-mahraganat-genre-music.
112 https://www.theguardian.com/environment/2020/jul/10/parts-of-england-could-run-out-of-water-within-20-years-warn-mps.
113 https://www.theriverstrust.org/2020/09/18/new-ea-water-quality-statistics-show-failure-at-a-national-scale/.
114 https://www.wwf.org.uk/future-of-UK-nature.
115 https://www.indexmundi.com/facts/indicators/ER.H2O.INTR.PC/rankings.
116 https://www.thetimes.co.uk/article/sunday-times-rich-list-earl-cadogan-family-wealth-vjc06w0r6.